GENERATIONS

Ned and Guy and René wove their webs round the three women, that night, it seemed to Marianne later looking back. There were other men, old and young – too old or too young – but those three were the knights. Marianne had found her dream material, Kate had summed up the field quickly and decided on her future course of action, and Sarah had felt pleasantly content. The ice had been broken for their season. Now they could grow up.

Also by June Barraclough
**available from Mandarin Paperbacks*

Heart of the Rose
Kindred Spirits
Rooks Nest
*A Time to Love

June Barraclough

GENERATIONS

A Mandarin Paperback
GENERATIONS

First published in Great Britain 1990
by William Heinemann Ltd
This edition published 1991
by Mandarin Paperbacks
an imprint of Reed Consumer Books Ltd
Michelin House, 81 Fulham Road, London SW3 6RB
and Auckland, Melbourne, Singapore and Toronto

Reprinted 1994

A CIP catalogue record for this title
is available from the British Library
ISBN 0 7493 0548 7

Printed and bound in Great Britain
by Cox & Wyman Ltd, Reading, Berks

Act One: Sowing

1867–1868

I

The sun, which had beamed mildly and fitfully all that April day, was just about to go down over London. Everywhere in the West End there was the bustle and business attendant upon the pursuit of pleasure. In the crowded slums children crawled in gutters and drunkards began their serious night's imbibing. Respectable folk were safely home; the suburbs were peaceful, lit up by the gas mantles of rooms in little villas or terrace houses. In Buckingham Palace the Queen, six years a widow, was engaged upon writing one of her long missives to her eldest daughter Princess Vicky, exiled by marriage to Prussia.

In one of the pleasantest parts of London, neither Mayfair smart nor suburban cosy, Madame Duplessis's Young Ladies had just had their supper and were preparing for the concert to be given that evening by selected students in the downstairs 'salon'. The finishing school run by Claire Duplessis was not the best known or the largest, but its small size and restricted intake enabled it to be patronised by old-fashioned county families as well as middle-class girls whose papas could (just) afford the fees for a few months' brushing up preparatory to entrance into society.

Three of the girls who were soon to leave Landon Place were sitting talking instead of tidying their rooms or practising their piano pieces. Indeed, of the three, only Marianne Amberson was to play that evening. The other two occupants of the shared attic room, Kate Mesure and Sarah Gibbs, were not regarded as musical, or at least had cunningly managed to convince Madame that this was the case.

Marianne was a slender girl of rather less than medium height, with her long golden-brown hair in an enormous plait. Her dark blue eyes looked out from a face that was by turns animated or anxious. She was at present trying to fix her hair up on her head and groaning at her own clumsiness.

'Let me do that,' drawled Kate, who was American and very chic.

Marianne submitted to an expert coiling and tucking, but held her head rather gingerly when it was done. 'I hope it doesn't fall down in the middle of my piece,' she sighed.

Sarah Gibbs, a tall, bosomy, fair-haired girl, laughed. 'Don't worry, dear. Kate's hair *never* falls down. She is an expert hairdresser.'

Kate, the darkest of the three, with the neatest figure and the most assured deportment, frowned slightly and then applied herself to her own toilette.

There was a silence while Sarah searched in a sewing basket and Marianne sat on her bed and opened a book.

'I wish I were a poor girl,' remarked Sarah then, threading a needle somewhat clumsily in order to sew on a slack, recalcitrant button.

'Why not let Queenie do that?' said Kate in reply. 'That is what she is paid for – to see to our clothes.'

'If I cannot sew on my own buttons,' sighed Sarah, 'what *am* I good for? Besides, it's nice and cosy up here, just the three of us. I hate the maids always coming in and out. It's worse than at home.'

'I agree,' said Marianne. 'Poor girls are allowed to do so much more than ladies – even if they have to sew on their own buttons and work in factories. Neither do they have to come out.'

'I am not a lady,' said Kate, looking critically at a mole on her face. 'Americans are never ladies however much money they have – not in England anyway.'

Marianne looked up from her book. In a way she envied Kate her independence, since the American girl's parents were dead, and she was to come out from her Uncle Adolphe's London house in Westminster Square. Sarah's parents, Sir Theodore and Lady Gibbs, were to take a relative's apartment for three months only in London, preferring their country life and only willing to exchange it for town for the sake of their daughter. Her own parents now lived in the country too, though they had once had an establishment in town and shown all the panache of *nouveaux riches*. They were now reduced to a mouldering manor in Oxfordshire since Marianne's father's business and investments had not turned out well in recent years. She was lucky to be coming out at all, she thought ironically. Kate too was still thinking about money.

'My uncle holds some of my money in trust for me for when I marry – or when I'm twenty-one – whichever is the earlier,' she said. She was determined that marriage would be the earlier event. 'And anyway, whether we are "ladies" or not , we have more freedom when we marry – not like poor girls.'

Marianne looked doubtful. Dreamily she moved over to the small looking-glass near the window. 'Mama and Papa wish me to marry in my first season,' she sighed, 'and I'm sure to disappoint them.'

Marianne's romantic ideas of 'love' were well known to the other two and Sarah smiled whilst Kate shrugged her elegant shoulders.

'You will be sure to meet an interesting "*parti*" when you come to Lake Hall Park,' said Sarah comfortingly. 'My mother is to invite my French cousin as well as all the usual county set for the pre-season house party.'

'Oh, Sarah, *you* will marry young, I am sure,' said Kate, plucking her eyebrows now and frowning.

Sarah Gibbs looked doubtful.

'Oh, you *will*,' pursued Kate. 'Your parents are the only real insiders in society – me being a Yankee and Marianne being the scion of business.' She preferred to be thought of as an outsider, believing that society was more malleable for the intelligent outsider than for those hemmed in by social prohibitions.

Sarah replied to her with a little frown.

'So much is expected of one, you know – not that anything is actually said – it's taken for granted. I'd rather live like you or Marianne.' Marianne gave a little moue of protest but Sarah went on, her cheeks flushing a little pink. 'Papa's sister will have lined up a few Honourables for me or a stray baronet or two, but they will terrify me, I know.' Sarah was shy, except among her friends, and would have preferred to live quietly in the country for the rest of her life, so long as her husband liked horses.

'Oh, Sarah, you are ideal breeding material,' said Kate vulgarly and Sarah blushed even more. Marianne thought, well, breeding *was* how you defined a lady, but you did not refer to her breeding capacities as though she were a mare.

She knew that her own parents were expecting her to carry off a glittering prize to redeem the family fortunes so refrained from commenting on what was to be expected of her. They were pleased that she had been invited by Lady Gibbs to a house party for the pre-season, and that Sarah's mother had agreed to present her. It would launch her once and for all. Marianne must not be intimidated by the county – the Ambersons still kept the slight insolence their earlier fortunes had endowed them with. Since she knew that she was regarded as pretty, she ought to be able to please them, but her real self, she thought, was unconnected with her pretty face and her conversational powers. Her parents would never understand anyway. Marianne had always been known for her 'unsuitable' feelings, her excessive sentimentality and her tendency to moon over curates. Now that she was nearly grown up they thought these attributes would have passed, particularly after nine months at Madame Duplessis's in South Kensington where curates were few and sentiments were curbed and the social butterfly was groomed to emerge from the schoolroom chrysalis.

She was silent, thinking about this, until Kate laid down her tweezers and said, 'We must go down. Thank goodness there are only three more days. I am so *bored*.'

Kate's ennuis were also well known. The three girls were a little set who kept themselves slightly apart from the twenty or so other boarders. They were known for their intelligence but also for their unconventional conversations, even if their behaviour was, on the whole, passable. Each in her different way was striking to look at; Kate with her gleaming raven's wing hair; Marianne with her romantic blue eyes and full mouth and Sarah with her long legs and rather pronounced bust, which was squeezed as far as possible into her 'black body' – her much disliked stays.

'Don't forget your music, dear,' said Sarah to Marianne. Marianne was a decent, if not outstanding, performer on the pianoforte. The audience was to be composed of the young ladies who still had another year to survive before coming out. The Principal's young ladies were trained in every way to be agreeable, and piano-playing and singing were part of the repertoire. It also included deportment, a detailed knowledge of etiquette, a little appreciation of flowers, paintings and bibelots, a little French, a little training in berlin work, French cuisine (not cooking, of course, but knowing what to order chef to cook and how to spell it on a menu), and – in order to appear extremely serious and to found good Christian homes – an extensive, if superficial, knowledge of selected parts of the Holy Bible. As a Roman Catholic Kate was not required to attend these classes. Kate however did all things well with the utmost outward coolness. Sarah followed her curriculum with some small detachment, but Marianne found it all a little beneath her notice and yearned for real learning.

'It is all so trivial,' she would flash out when bored by the needlework or the etiquette. 'I should like to learn Latin and Greek and to read in our great English poets.' As it was, there was hardly time to undertake any serious reading at the school and she looked forward to the vacations and to returning to the circulating library of the little town near which her parents now lived. Her great ambition was to hear Mr Dickens read. The library at Landon Place was a dull one. Only a select few of Mr Dickens's works were allowed and none of Mr Thackeray's. The girls read other more lurid fiction, of course, when they could come upon it, and nurtured their young imaginations upon the lives of lost girls and wicked women – and Kate had purloined various novels with yellow covers from France, written in a language she spoke well for she had lived on the Continent.

Sarah had access to a grander library at home: old leather-bound volumes gathered over the centuries which were scarcely ever opened by any of her relatives. Her father, Sir Theodore Gibbs, occasionally wandered there when he was not hunting or shooting or fishing, and Sarah's French uncle, the husband of her mother's half sister, a Monsieur Paul Boissier, stayed reading before the fire on his infrequent visits to Lake Hall Park. Otherwise the books were just items of furniture, handsome to behold. She had however promised Marianne the run of their library when she came to stay and Marianne looked forward to discovering tomes from which she might transfer some knowledge of the world into her own head while Sarah went riding. Kate too was not unconversant with book knowledge, but in her case she needed books to tell her how to live in an agreeable way, not to unlock the secrets of the universe, for which she cared not a fig. Since she would be an heiress, she intended to enjoy herself, to travel, to own many beautiful objects and one day perhaps to run her own salon. Not in stuffy old England, but in Florence or Paris or Rome – or even one day perhaps back home in Boston. Women must look after themselves; she did not think men were up to it, though a husband would be *de rigueur*, at least at first in order to establish oneself.

As they went down the stairs, Marianne, far from concentrating on her piano piece, was wondering if one day soon her ideal man would be talking to her of love and literature – she did so want to meet literary men.

Surprisingly, she had no nerves when she had to perform in public – her fears were for more intangible things, or for water or heights or horses. She enjoyed showing off a little, provided she knew her stuff – a poem to be recited or a soulful sonatina to be interpreted. When she could forget herself her other self emerged. It was her turn that night, after twenty minutes or so of a plump girl singing Mendelssohn. The three friends sat together at the back of the room waiting for Marianne's call which the other two would applaud whether she played well or no.

Kate looked round the salon at the two (in her opinion) frumpish ladies who ran the establishment, the nervous music master (an anaemic individual), and the twenty or so young ladies ranged on little gold chairs and dressed in their best crinolines, looking for all the world like little birds tethered to the ground by their feathers. Madame Duplessis's establishment was not very smart – no, definitely not, but she thought that she and Marianne and Sarah lent it *éclat*. Not that Marianne was smart, but she had an otherworldly air which was at least interesting, and Sarah looked what she was – the healthy scion of a landed family, polite and large and easygoing and blessed with taste even if she did not know it – English taste anyway.

It was Marianne's turn at last and she went up to the piano quite confidently. She knew she was not a very good pianist but no one else, apart from Mr Young (the anaemic gentleman), did, and she would play her best and even extract a little 'soul' from the music which might make up for any wrong notes. She played an easy *étude* as the gas candles flickered on. It seemed to speak of another world, one where feeling was more important than outward show, though it needed the application of 'style' to be evident. She was puzzling on this paradox even as she played, but managed to get through the music and even draw a smile from Mr Young. The young ladies clapped dutifully when she finished, though there were a few polite yawns suppressed behind fans. She went back to her chair at the back, pleasantly gratified. Sarah was clapping with vigour and Kate poked her with her fan and mimed rhapsody. 'Terrific – first rate, dear,' said Sarah.

All three felt content as the next item arrived, a song from a young lady whose father was a City banker and who paid for her to have extra music lessons. They did not compensate for a complete lack of pitch. It was a pity that the parents of the young ladies could not be present, thought Kate, for at least it would show them the folly of spending their money on a girl of no discernible talent. But music would be needed when gentlemen were present once the girls were out – it was a snare to attract them into offers of marriage.

They were all thinking that it hardly seemed possible that in a few weeks they would have left Madame Duplessis and the boredoms and inanities of the finishing school. They were all over eighteen – quite old to be launching themselves for the first time upon society. Various events had conspired to make them miss their season the year before. Sarah had had a hunting fall, Marianne had had scarlet fever, caught from one of her younger sisters, and Kate had been abroad for six months in Italy with her American guardians.

At this time of their lives Sarah was possibly the most content of the three. She found happiness in the contact of her skin with air and water and longed to be back with Star, her chestnut mare; she would have been happy to relinquish any thoughts of marriage or social success if it had not been understood that she must make some effort soon to conform. The projected house party, given in the 'little season' in April, as a *bonne bouche* for later, was to be the scene of her first attempts to show off her marriageability. But she comforted herself that both her friends would be present: their private début was to be at this party, whatever the London season would do for them later. She could fit some riding in and perhaps some archery. Her large body, so ungainly in fashionable clothes, gave her no trouble if she was doing what she enjoyed. Sarah was not too reflective, rather lazy in fact except for outdoor sports, a little naive in matters in which she took no interest, but unsnobbish and tender-hearted. Her lack of reflectiveness was not the result of any shallowness but a genuine deficiency of interest in herself. She was at a deeper level happily in tune with what Marianne called The Universe.

Marianne was thinking about her friends as they sat on in the salon awaiting the next solo. Neither Kate nor Sarah was the anxious individual she was herself, however much she concealed it in public. Kate had listened without shock to Marianne's frequent expositions of the problems of the reality of the self or her tortuous attempts to believe in a good God, but also without much interest. Kate, thought Marianne, was fused in mind and body – she did not give much time to agonising over ideas, though she found most intellectual work easy. Sarah would smile and say, 'Don't worry, Marianne – I expect some people have always worried over that sort of thing.' 'You are too anxious, Marianne,' said Kate. 'You should forget all these worries since there is no answer to them and never has been. You must learn to enjoy yourself.' Marianne knew that the spirit which joined to her when she was listening to music or even sometimes playing herself, or when she was reading or painting, deserted her when the real business of living obtruded. Love, of course, that was different, and was surely more like getting in touch with the Universe – but there seemed to be so many obstacles to love. Kate was tuning herself for an attack upon society, one which she was wise enough to know must be concealed under a veneer of cool elegance. Marianne found such elegance beyond her and neither could she dissimulate when it came to feelings. Marianne with her anxiety and her physical timidity yet concealed a rebellion whose only public manifestation was a certain hectic vivacity. Strangely enough people seemed to like this side of her and no one would have guessed she was in reality as shy as Sarah and quite as calculating as Kate, but without the hardness to carry it through.

When she was a small child and her mother was preoccupied with her little brothers and sisters, she had been given over to the charge of a Nurse Wilkins, a sadistic woman who delighted in humiliating her. She had hated Nurse Wilkins with a fierce singleness of mind unusual for a timid child. Nurse Wilkins had aroused Marianne's spirit of self-protectiveness. One day Marianne had turned upon her when she was being beaten for insubordination (not eating her cold pie which made her feel sick). Marianne had then actually bitten the woman and been banished from the nursery for a week with a fever and hallucinations and terrible nightmares. When she recovered, Nurse Wilkins had gone. Her father had had something to do with that, she knew. Papa had said she might have been a bad girl but he understood why she had been so naughty. She had felt deliciously happy then, for Papa was often away and it had only been by chance that he had returned home the night of the nursery battle when Marianne had been removed, still kicking and screaming and sobbing from the *scène de combat*. Somehow or other Papa had got round Mama, and Wilkins had gone.

Marianne did not like her Mama very much, though she loved her. Mama was always having a new baby and had no time for Marianne. The other servants had liked Nurse Wilkins no more than Marianne did and things went on better after her dismissal. But Marianne felt ever afterwards a terrible guilt. One day, she was sure, she would come upon Nurse Wilkins when she was alone, with no Papa to stick up for her, and Wilkins would kill her this time. Yet, in spite of her passionate reaction to the injustice of the beating and the horrible nausea that had followed her self-defence, she had never shown quite such spirit again. When Papa took the family to Ramsgate and they bathed from the machine, it was she who most feared the water. When Mama urged her to learn to ride her pony (they had moved to a new house in Hertfordshire), Marianne had never enjoyed the lessons. When the boys had challenged her to climb up trees after them, she had been dizzy and not able to do with her body what her mind longed to do. She knew she was a physical coward – but hoped she was not a moral one and when she saw men snaring birds one summer she had not been afraid to speak her mind. She had hated the men with all the passion with which she had once hated Nurse Wilkins and she had hated her brother when he tore off the wings of flies. But why was she sometimes so angry and yet so cowardly when it came to doing anything physically dangerous? It was not her idea of herself and she was bitterly ashamed. She was a worried child who had found as she grew up that she could compensate for this by the expressivity of her conversation and her infectious enthusiasms. She saw her real self as being in her mind – a daring imaginative self of which only her two dearest female friends were aware. Sarah was kind and unshockable but was genuinely angry when Marianne told them one day of the Nurse Wilkins incident and Kate said: 'Good for you, Amberson!'

As the girls trudged back to their top floor eyrie after that evening's concert, the talk was of the forthcoming house party at Lake Hall Park once the shackles of school were thrown off.

'It will be too delicious,' said Kate. 'At last I shall see how the

English really live. Sarah, dear – I hope you have warned your Mama about my Yankee manners?'

'I have told Mama and Papa all about you both,' said Sarah. 'They are looking forward to meeting my friends, but you know, they are not great entertainers nor fashionable – they are just doing what Mama calls her social duty, so you will enliven proceedings for them in a way I cannot.'

'Let's plan it all in the morning,' said Marianne. 'I want to think before I go to sleep. Oh, I shall be glad to have my own room again.' Then she realised this might be taken as a slur on her companions, so added, 'I mean, I'm always frightened of keeping you awake with my candle . . . when I read – '

'My dear Marianne,' said Kate, pulling off her slippers, 'we know what you mean – won't it be fine to have a little space of one's own – we all feel the same.'

'And as for keeping *me* awake,' added Sarah, 'nothing ever stops me falling asleep. Papa says I could sleep on a horse.'

'I'm not going to read tonight,' said Marianne, pulling out the ribbons from her long hair. 'How glad I shall be when all this lot is up.'

'When your hair is up and we are out,' said Kate. 'Mesdemoiselles – there will be such a transformation!'

2

It was not to be a large party; Sir Theodore could not abide large parties, but it was to be a select one, and he left the management of the affair to his wife. Lady Gibbs, although she looked vague, could be quite competent and practical if necessary, and rather enjoyed the planning and execution of their attempt to do their best for their daughter. Sarah was not much help, preferring to ride round the estate or to accompany her father in his pursuits when possible. Her tallness and plumpness were metamorphosed into the look of a young Juno when she wore her riding attire. If only she could be wooed in her outdoor clothes, thought her mother, there might be hope for her. In her evening crinolines and flowers she would look more like a placid giant, her fair locks escaping from the chignon in which her long hair would be braided. It would be the sign that she was out, but would not please her, except that it was more practical for riding.

There were several county couples whom the Gibbses always invited to their house parties, old friends of both husband and wife. Sir Theodore had married his Lady Penelope rather late in life and she had not been young at the time. From this their daughter took comfort – perhaps they would let her stay on at home till she was thirty or so and then she might marry some kindly older man with similar interests to herself.

'I wonder if we should invite the French connection,' said Sarah's mother to her husband. The guest list had been drawn up some weeks before, but several had cried off. One set of friends was off to Italy for Easter; another had had a death in the family. Lady Gibbs's half-sister's only son, René Boissier, was a good-looking, if rather enigmatic, fellow of twenty-nine. He had studied law but seemed nevertheless to follow his father in a certain admiration for opera girls. Sarah had only once met René, her first cousin, years before, but he had impressed her then as being handsome and kind. He had taken more notice of her than either of her two English cousins had done. He had seemed old for his age and much more sophisticated than English boys of eighteen. This René Boissier had just written to say he would be over in England on an extended tour with a friend and so Lady Gibbs felt obliged to invite him to her party. 'We so rarely see any relatives of mine,' she said to her husband. 'He might take a fancy to an English gel – Susan would be pleased to renew her English connections, I'm sure.'

'So long as he can speak English,' was her husband's reply.

'Why, of course – you remember Sukey was most insistent he spoke English to her. When he was a little boy he had a charming way with our language – after all, he *is* half an Englishman.'

Her list was finally drawn up and at the last moment there was added to it a Monsieur Guy Demaine, René's travelling companion whom he implored his aunt to receive. Then there was young Ned Mortimer, the son of Sir Theodore's best friend 'Tubby' Mortimer with whom he had been at school before the latter retired to his estate in the North of England and to the founding of a manufactory.

''Twill only be a short stay,' she said. 'Afterwards we shall have the trouble of launching Sarah in London – so we might as well enjoy it and invite the young people who will enliven things.' For besides these new guests, there were several families of girls, the grandest of whom were the daughters of a second son from Mundon Park, brother of Lord Maltravers. They would stay with their Mamas and Papas and be chaperoned by several ancient aunts from their own and other families of the district.

At last the invitations had been dispatched and the replies received, by which time Sarah had arrived home from Madame Duplessis and was eagerly awaiting the arrival of Kate and Marianne.

Lake Hall Park had begun as a small manor house in the seventeenth century, built out of Cotswold stone. But the next century had seen the addition of a whole new wing which now faced the front drive, presenting a tall, pillared portico to visitors as they arrived. The gardens and woods and fields had been landscaped and a small stream dammed to make an ornamental lake, and a vast stable block had been erected at the back of the original house, with a new courtyard. The newer wing boasted a long gallery extending over the front elevation and looking out over the gardens and paths and lake to the left. Sir Theodore was at that time finding it difficult to manage the whole estate with profit, and he had recently had to sell his Jamaican plantation. As yet, however, Lake Hall still looked fairly grand and he had not yet sold off any of his fields to neighbouring farmers.

It was a warm April and the primroses were out in the woods and cowslips in the fields. Sarah woke to the sound of the cuckoo the morning her friends were to arrive and dressed hastily. There would only be two days before the other guests arrived on the Friday and she intended to show them the place and for them to enjoy each other's company before the disagreeable business of husband-hunting and dancing began.

Kate arrived first that afternoon from a train that stopped in the nearby town, accompanied by an elderly aunt who was eager to sample the delights of an English country house. 'And there will be archery and croquet – and riding – thank goodness you learned to ride in the States – ' she was saying, all of a quiver with anticipation. Kate found her very boring but tried to conceal her feelings – it was bad enough to be chaperoned, even worse to have such a gushing example of American spinsterhood continually by her side. But Aunt Carrie did not ride and Kate had decided that therein would lie her own salvation. The one place one could be alone or accompanied by an agreeable man with no prying eyes would be while out riding.

'Promise me you won't feel it necessary to follow me,' she said to Sarah as they sat unpacking Kate's wardrobe with Sarah's mother's maid in attendance. Aunt Carrie was already in the drawing-room sampling tea and cucumber sandwiches and about to be taken on a tour of the conservatories by Sir Theodore.

'You can be as free as air here,' replied Sarah 'My parents are very old-fashioned. I shall probably escape myself on Star.'

'When is Marianne arriving?' asked Kate.

'This evening – they are bringing her over in her father's carriage – poor Marianne, she will hate the riding.'

'Well, let her play croquet with the clergymen and spinsters,' said Kate.

The great event was the ball to be held on the Saturday evening and for this the long gallery of the Hall had been transformed by Lady Gibbs and her servants into an indoor conservatory. The

Gibbses did not live in great style, but old memories of dances persisted in their heads and all was to be pleasant and enjoyable without being in any way ostentatious.

'Pretty place you've got,' said Kate as she went round the stables after tea with her friend. 'When are the *partis* arriving? Tell me all – who is to be here?

'Well, there are all my old friends from childhood who live near – mostly Mama's and Papa's friends' children and one or two relatives and the curate and – '

'You don't make it sound much fun!' objected Kate.

'Oh, there will be friends of Papa's from the county and a politician or two, I expect, and some Army men and my French relatives – at least only one of them – but he is bringing a friend. And there is a nice young man called Ned Mortimer whose father was Papa's greatest friend when they were young – '

'So I can practise my French?' interrupted Kate. 'Can we sit down in the conservatory and chat – I find walking up and down rather distracting.'

'Why, of course – I always forget that young ladies are usually more pleased to sit than walk,' said Sarah. 'I never get tired myself – it must be the country air – sleepy sometimes, but not tired.' She led the way into the warm, peaty-smelling glasshouses and the two girls found a bench which Kate reclined on after sniffing the forced flowers that were the gardener's pride and joy.

'Tell me about the Frenchmen.'

'Well, there is my cousin, René Boissier, whom I haven't seen for ages – but he is a great rattle, I believe, and rather a sweet person – at least when I was little he was nice to me. He is nearly thirty, I believe. And he is bringing a friend, a Monsieur Guy Demaine, Mama says, she has done all the arranging.

'Fortunately my aunt does not speak French,' murmured Kate with a slight smile.

'Oh – do you hear carriage wheels? It must be Marianne!' said Sarah.

They had a view of the long path that wound in from the lodge to the Hall and a small carriage was just arriving. 'Her Mama will want to stay for a chat, I expect. Come along – let us go and rescue her.'

Kate got up rather half-heartedly, absent-mindedly crumpling the leaf of a redcurrant bush between her fingers as she followed Sarah down the path.

Soon excited voices and the yapping of dogs mingled at the bottom of the *perron* that led to the Georgian part of the house. Marianne was finally extricated from the carriage and a rather dark, dumpy lady introduced as Marianne's Mama. Soon, however, as Sarah had predicted, she was borne off by Lady Gibbs to partake of some refreshment before her return home.

'I wanted to come with Tom, my brother, but they would not hear of it. I expect Mama will be discussing the chaperonage with your mother now.'

'Mama has never insisted on too much fuss, said Sarah, 'and has persuaded various elderly relatives to see to that side of things. Now, Marianne, let me look at you. You do look smart.'

Marianne, who had been given a rather small sum to spend on her attire, owing to the straitened circumstances of Mr Amberson, looked gratified. She was wearing a dark green costume which Kate privately thought too old for her, and a small round-brimmed hat with a curly feather. 'I hope it will not be hot for I have forgotten my straw bonnet and my chipstraw hat,' she said.

'Oh, never mind – you can borrow one of mine,' said Sarah.

'What is your ball dress like?' asked Kate curiously. It had been the subject of many fittings and consultations and agonising. None of the girls had, as yet, an income ample enough to provide a large wardrobe. Kate looked forward to the time she would inherit her fortune and could buy as many ball gowns and costumes as she wanted. Meanwhile she had had a French dressmaker confect her something in the very latest fashion from oyster silk. With it she would wear her own inherited jewels, the pearls which set off her dark complexion. Thus dressed and bejewelled, she would be the very picture of young female innocence, but the knowing eye would remark a certain *tournure* in the cut and a certain way with a train that were indisputably Parisian.

'It is blue,' replied Marianne to Kate's question. 'Sky-blue with a dashing crinoline and I shall wear blue flowers and carry a blue fan.'

Blue was decidedly not in fashion, thought Kate, but Marianne would not know that.

'And you Sarah? How did it turn out? Did they manage the frills? You said there was some difficulty.'

Sarah had been made a white dress with pink rosebuds in the weave that was really very pretty, but rather childish for her statuesque figure. It was, however, fresh and dainty and her father said she looked a picture in it – 'Just like your dear mother,' – when she had advanced for inspection.

'Of course, Marianne, you will have a white dress too,' said Kate. 'Apart from your presentation one.'

'Yes – they are buying one in Birmingham.'

'*Buying* one – in *Birmingham*?'

'Yes – there is a shop – Papa had ordered it for me. We thought for here the blue would be better – '

'You will look lovely in blue,' said the generous Sarah. And I hate white – even though we must wear it for being presented. Vestal Virgins, don't they call them? White doesn't suit me either.'

'White roses though,' said Marianne dreamily. 'I must have white roses – '

The rest of the evening was taken up with discussion of guests and toilettes and gossip about other school companions. 'Thank goodness Landon Place is over,' said Kate. '*En avant mes amies*, I shall drink to our triple attractions at your father's table tonight, and if your father toasts the Queen, that is what *I* shall drink to.' Sarah looked a little shocked, but Marianne laughed. If only she could be as confident and glamorous as Kate, who would never wear a blue dress and who would obviously be the cynosure of masculine interest.

Later, when they looked back on that long weekend at Lake Hall Park all three of the girls were to feel that their real coming out had been then, not the feverish dash and rush or stately boredom of the weeks that followed in London. The features and figures of three young men were to dominate their waking and dreaming selves after that initial entrance into the world beyond the schoolroom.

Ned Mortimer, Guy Demaine and Sarah's cousin René Boissier were the three cavaliers at Sarah's ball among whom their favours were distributed. Sarah was happy to see Kate slide into conversation in between dances with Cousin René, who was certainly a handsome young man, quiet that evening, and who seemed to have none of the Frenchman's mockery of English habits or sneering at English food and drink. He had an unmistakable air of masculine authority about him, not the same sort as her Papa or any man she had ever met, more the authority of a man who enjoyed being himself and wanted his partners to enjoy it too. She had been tongue-tied with him at first when he had taken the first dance with her as her nearest young male relative. But he had not seemed to notice and said only, 'You still enjoy the riding then, Sarah?' The way he said 'Sarah' with a French 'R' made her shiver. It was like being a little girl again, chosen by him for a forfeit. When he danced with Kate, she noticed that he looked quite different, more aware of Kate as a woman. Kate looked dazzling, but had not seemed to show any particular emotion when for the next dance he claimed Sarah again, and Kate, in her turn, was trotted up and down by Ned Mortimer who was a large, benign, kindly man of about thirty. Then Ned danced a *valse* with Marianne, the most daring of all dances in polite society. Marianne seemed distracted and stumbled a little and kept excusing herself. For a musical girl she had a strange lack of rhythm in her feet. Ned did not appear to mind; he was a taciturn man but occasionally he listened to something she said with an expression of real interest in his eyes under their bushy Northern eyebrows. Marianne relaxed once she could sit down and fan herself, or pretend to. But she was not allowed peace for long; René Boissier's friend Guy Demaine claimed her shortly afterwards for a polka and she quite came to life again. Kate, dancing now with Ned, had her eye on them both and thought that Guy Demaine was the most handsome man she had ever seen, far handsomer than Sarah's cousin René. She wondered idly if he was a fortune hunter. His chin was perhaps rather weak?

Sarah was back dancing with her cousin. They were the two tallest people in the crowd of young men and women. Her fairness set off his dark, neat features. She realised that he was an exceptionally good dancer – that was why she felt she herself was at last dancing well. Marianne's next dance was again with Guy Demaine. She had better be careful – the Mamas and chaperones would be on the look-out for any excessive partiality shown by any young lady to her partner. But Marianne did not appear to care; she was looking in the eyes of the young Frenchman in a way Sarah remembered from when she read Byron to them in the dormitory or enthused over one of her literary heroes.

'All enjoying themselves,' said Sir Theodore to his wife. 'Sarah looks very pretty, Lady Gibbs.' He still called his wife by her title in the old style, and indeed looked himself like some survival from forty or more years before when he had graced the gaming tables of St James's in the reign of George the Fourth. Lady Gibbs smiled at him composedly. 'What about the Frenchies eh?' he asked with a lift of his eyebrows and a hearty laugh.

'They seem quite taken with Sarah's friends,' she replied. 'At least M. Demaine does. My nephew has been dancing with your daughter – she looks handsome tonight, Theo.'

'Not as handsome as you,' he replied gallantly. But she knew it was not true, that her husband was very fond of his little daughter, now such a big girl. Sir Theodore looked round for the son of his old friend. He hoped Ned Mortimer might take a fancy to Sarah – they seemed well-suited – but he espied him now dancing an old country dance that he himself had asked for specially, with Miss Kate Mesure. She looked a sly young puss, he thought, not like his Sarah. Ned was looking quite unmanned. Sarah's other friend, Marianne, was sitting this one out next to one of the chaperones, with René's friend paying court to her. He must stop gawping and see the next lot of champagne was uncorked to drink to the Queen, God bless her.

Marianne was feeling rather unreal and, for once, unaware of her nerves, her fears, herself – blessedly unselfconscious. Guy Demaine talked as she had never imagined a young man could talk to a girl – teasing, flattering, making her laugh. He was charming and he listened to her with every appearance of interest. His accented English was charming too. He danced quite well, was deft and practised in his movements, and above all had nice dry hands that steered her when necessary and lingered just long enough in hers. He seemed used to women. Well, he was French, wasn't he? This evening, the first evening of her womanhood, he was also, she sensed, making a special effort to attract her. Heavens above, she hoped he did not think she was a rich girl, an heiress like Kate. He even seemed to take an interest in some of the opinions which she thrust rather quickly and breathlessly at him. Yet he seemed, as well as paying her the compliment of listening to her, to be about to break into a smile. Perhaps after all he did not take her seriously. She had not time to ask herself though for they were off dancing again.

'You have beautiful eyes,' he said during this dance and she nearly swooned with excitement. No man had ever said that to her before. She must be careful. But she did not seem to care tonight. Certainly many girls would like to be dancing with him, and yet he had chosen her. She wondered why, was too genuinely modest to think that he could find her pretty. He, in fact, liked the vivacity which she was well able to pour out. One little part of her though was saying; 'It's all a game. He will never know the real you.' Perhaps learning to live in society was learning not to take oneself or others too seriously, Marianne thought. The façade might then become the truth. She wished she knew him, that they could really get to know each other.

'Tell me about your work,' she said.

'I work in the same chambers as René – he and I are in the law – and we invest too on the Bourse and try to make money.'

'My father is a businessman,' she said without coyness.

'I paint a little too,' he added. Then, 'I am not a nobleman like your friend's father.' He gestured to Sir Theodore who was now seated with his lady in a room that led out of the ballroom.

Was Sarah's father an aristocrat? He had a title certainly, but that did not mean he led the life of the charmed top families. They were more like old country gentry, she supposed – the sort of gentry her own parents did not quite belong to, though at one time they had had more money than the Gibbs family.

'We are professional gentlemen,' said Guy. 'Or is that wrong?'

'You are either a professional man or you make a profession of being a gentleman,' she replied.

'Ah, that is witty. You are a witty girl, Miss Marianne.'

Oh, she was enjoying herself. And the champagne helped. He brought her back to her chaperone and poured a glass for all three of them.

'Tell me,' he asked as they sat sipping the ice-cold drink which seemed to bubble inside her yet melt all her cares away. Your friend, Miss Mesure, she speaks good French, which is unusual for an American lady. Is she your own age? She looks older than you.'

'Yes, Kate has lived in France, and we are both eighteen, and Sarah too.'

'I would not have thought it,' he said. 'You, yes, in your blue dress you are the spirit of eighteen, but Sarah, she is already a matron, I think. And Miss Kate, she is a woman of the world.'

'Oh, yes, she is,' said Marianne ingenuously. 'Kate is an heiress. Her parents are dead, but her father was a rich prospector, I believe. She looks beautiful tonight, does she not?' she added.

He smiled. 'I see young Mr Ned come to ask you for a dance,' he said. 'Promise you will have the last one with me.' She promised and he was gone and Ned Mortimer was sitting beside her, so solid and English, not the sort of man who would find it easy to pay compliments.

Ned and Guy and René wove their webs round the three women, that night, it seemed to Marianne later looking back. There were other men, old and young – too old or too young – but those three were the knights. Marianne had found her dream material, Kate had summed up the field quickly and decided on her future course of action, and Sarah had felt pleasantly content. The ice had been broken for their season. Now they could grow up.

3

Everyone got up late the next morning, for the dance had not finished until two o'clock. A chill spring dawn had brightened the sky and the servants had been up for many hours before Sarah stirred. She dressed quickly and went out to the stables after swallowing a cup of hot coffee left with a meal in the breakfast room for those revellers who were staying overnight. As it was Sunday her parents would be in church. They would expect her to go herself in the evening – she was glad they had not insisted on anyone but themselves setting off for Matins. No one else seemed to be up, and the doors of the long gallery were closed as she passed it on her way down the back stairs to the courtyard. Only Perkins, one of the housemaids, whisked away round a corner of the staircase before her. The servants would also all attend church in the evening. Otherwise Sunday was hardly a day of rest for them. Somebody had to cook and lay fires and tidy up, even though it was the Lord's Day.

Sir Theodore went to church across the park out of habit; he would have been hard put to define his own beliefs. His set had never been sanctimonious like those damned Evangelicals. But the Church of England, bless her, was part and parcel of being an Englishman. He did not go so far as to state that God was an Englishman, but the Queen of England was one of His representatives on earth. On Sundays he enjoyed the drive across his land and the sight of the villagers in their best clothes. Not that all the villagers were churchgoers: some were too tired by Sunday; others were indifferent. But they had their needs looked after by Lady Gibbs as the Squire's wife; and the Rector, a gentleman, was always welcome at the Squire's table. On the whole the alliance between church and Hall worked well, and Sir Theo looked forward to his daughter's being married one day in 'his' little Norman church with its new additions and embellishments.

Sarah went to Star, her noble old head poking out of the stable door and a benevolent look in her eye. She gave the horse her favourite carrot tops and stayed for a moment by her, arms over the half door and cheek against the warm, brown flesh. 'I danced a lot,' she murmured, tracing the white star on the horse's forehead with a long finger, 'but I'd rather be with you. Promise to behave if I bring some young men out to see you before luncheon.' She had told Ned Mortimer he would be welcome to inspect the stables and they might go out for a ride on Monday perhaps with Kate and the French cousin. She had better go in and see the girls were woken up. She well knew how long Marianne would sleep if she were left undisturbed.

On her way back she came across Kate already dressed in pale grey and seated on a rustic bench.

'Why, have you had no breakfast?' she exclaimed.

'I'm not hungry,' said Kate. 'I believe your Frenchmen are eating breakfast. Marianne is having her hair dressed by your maid. My feet ache. These damnable slippers.' She stretched out her neat, slim feet from under her wide skirt. 'Have you been to the stables?'

'To see Star – and tell her about the dance. Did you enjoy it?'

'It went off very well, you may be assured. Your Mama must be pleased. And now we can, I hope, settle down to a calmer few days.'

'I've never known you, Kate, want a "calm few days",' observed Sarah with surprise.

'Oh, I'm looking forward to practising my French again,' Kate replied with a smile. 'Your cousin is charming – and his friend.'

They were interrupted by a footfall and Ned Mortimer came into the walled garden from the carriage walk side. He bowed when he saw the two of them. Sarah blushed.

'Do join us, sir,' she said. 'That is, if you have had breakfast?'

'And if you don't feel you should run for my aunt to act as chaperone,' added Kate with a wicked smile.

He looked a little nervous, but took the remark in good part. 'I breakfasted early, couldn't sleep, all that excitement. Dancing goes to the head, I believe,' he said, sitting down solidly by their sides.

Kate smiled in silence and Sarah felt obliged to make conversation.

'We have croquet and archery round in the New Garden – at

least tomorrow it will all be arranged, I believe. But I thought you might like a canter in the park in the morning. Papa would enjoy your company – Mama no longer rides.'

'And do you ride, Miss Kate?' asked Ned, greatly daring. He had never met anyone quite like Miss Kate Mesure before. She seemed altogether too poised, clever and stylish for a girl of eighteen summers. But, of course, he was a country bumpkin and could not judge.

Ned's father's new manufacturing business had made profits which, shrewdly invested, had bought an old farm and a house in the North Country far from these Southern gentilities. He had even begun to talk Northern and Ned himself had a slight Northern accent. Ned had been brought up as a gentleman farmer with income generated from manufacturing. He felt it was an uneasy compromise and had recently entered the firm to learn the trade.

'Oh, I'm a complete novice,' said Kate. 'I do, but badly. Provided the horse is obedient and I don't have to hunt.'

'A *com*-plete novice,' he echoed. 'Oh, I can't believe that. I like a bit of trekking up in the hills. It's all a bit civilised down here.' He gestured vaguely round the park. 'Beautiful, of course,' he said, turning to Sarah. 'And where is the other young lady – Miss Marianne? Last night she promised me a book. Full of it she was – I promised to read it!'

'Oh, Marianne will still be indoors, I suppose,' said Kate. But just then she appeared, clutching her head and pushing all the unaccustomed ringlets back into her chignon. She started when she saw Ned and made as if to turn. Ned bowed though and she came up. She could not mention the agonies of hairdressing to him. A pity. She wanted a long chat with her friends before luncheon. Ned sensed that he might be *de trop* and muttered that he must see Demaine about a carriage into Oxford the following day.

'Oh, Guy – Mr Demaine, is not leaving tomorrow surely?' exclaimed Marianne before she could stop herself.

Kate exchanged a raised eyebrow with Sarah. 'I believe they are to look round the University but return in the evening and stay two more days. I hope *you* will stay a little longer,' said Sarah politely to Ned.

'I have to visit town on my father's business,' said Ned. 'But I look forward to a ride, Miss Sarah. Will you be going to church this evening?' he asked but his eyes were on Kate.

Sarah answered, 'I expect we shall all go – I'm afraid we were not up in time for Matins today.' Unlike many families of their rank the Gibbses did not have daily family prayers.

Ned stood awkwardly a moment then bowed again and took his leave. When he had disappeared they all began to giggle. 'He is perfectly *sweet*,' said Kate. 'I do declare, Sarah, he would be the perfect husband for you!'

Sarah blushed. 'Nonsense, he couldn't take his eyes off you two. Did he pay you many compliments?'

'I didn't notice,' said Marianne. 'He is very nice, but a little shy. He doesn't dance as well as the Frenchmen. Oh, I *did* like Monsieur Demaine! He is breakfasting – perhaps if I breakfast now . . .'

'They will all have finished,' Kate said quickly. 'I thought M. René charming – rather like a Frenchman I met last year in my travels. You liked M. Demaine then? It's true, he is excessively handsome.'

Marianne said nothing, but continued to look 'spooney.'

'He is handsome,' Sarah agreed. 'But I believe he has no fortune. Cousin René told me he wanted to be a painter – he doesn't like business.'

'And Mr Ned, I suppose, *does* like business?' asked Kate.

'Yes, he is comfortably off, but prefers to visit his father's factories in Manchester and return at the end of the week to Westmorland where he enjoys the country.'

'Come on, Marianne, you will be able to charm Monsieur Demaine at luncheon,' said Kate. 'And you can hold his brushes for him if he wants to paint the park.'

Marianne gave a nervous laugh. She had had a slight headache, but it was wearing off now and she felt suddenly full of energy. She had lain awake for what seemed hours thinking of the handsome Guy and only fallen asleep when the light was already seeping through a crack in the blinds.

Kate thought, now there's a situation. Mr Ned is obviously smitten with me, and Marianne will have eyes for no one but Monsieur Demaine. But Ned would suit Sarah down to the ground. I wonder whom *I* like best? Monsieur René would be the most suitable, but I have the feeling he will marry someone who is not too much of a challenge. Ned would be nice to 'improve', he is full of sterling qualities, I'm sure. But Marianne is right – Guy is the most interesting, in spite of his chin. I wonder with whom Monsieur Guy would prefer to spend his time. He is used to light women – that much is obvious – but his looks will probably gain him an entrée into artistic and aristocratic circles. She said nothing of this of course. She had had the impression that Monsieur Demaine's glance had also rested quite often upon herself.

Marianne and Kate were to be presented at Court through Lady Gibbs's good offices. Sarah was to be presented by her mother. Kate had already come out among her own American circle, but the house party at Sir Theodore's was her chance to shine among the English. Since Marianne also had been accorded the honour of appearing at a pre-season party in the company of the Gibbses, she could appear out in the new and sadly provincial business circles of her parents, put her hair up and ape the manners of her betters. The Gibbses might not be very rich, but they were well connected and did not need the immersion in smart society that lesser *nouveaux riches* families or greater more noble families took as a matter of extreme importance. That Lady Gibbs's sister had married a Frenchman and no longer lived in England, that Lady Gibbs herself was a little unconventional, all added to a reputation for slight eccentricity. Sarah would stay at Sir Theo's sister's while in town and be launched from there. This duty done and, it was to be hoped, a suitor arising from it, her parents could then retire thankfully once more back to Lake Hall Park. Marianne's parents would, a few years back, have had her presented themselves if she had been old enough; now that their fortunes had declined (though ever expected to take up again) they were content she could appear in the house of such a respectably old and stable family as the Gibbses, even if it were not of the smartest. At Lake Hall proprieties were, of course, on the whole observed: the girls were chaperoned by a married lady wherever they went – as far as Lady Gibbs knew and as far as *her* arrangements went. But there are times and places where even the lightest of chaperonage may be shrugged off. One such occasion was the smaller more domestic party that followed the ball on the Saturday and the pleasant Sunday chats and the dead Monday when the young men visited Oxford. It was Kate who suggested a game of Hunt the Slipper on the Tuesday evening after the party had played charades in the drawing-room and Sir Theo and his lady had left the younger members of the party and adjourned to a smaller drawing-room to talk or snooze.

There were about fifteen young people left, some sitting stiffly on upright chairs, others, including Kate Mesure, draped on a sofa near the fireplace or standing in little groups. There had been a song from a young lady before the older members of the party left the room and Kate felt she could not put up with any more of that. Besides, Marianne was casting soulful glances at Guy Demaine, who seemed oblivious, and Sarah was trying not to look bored but wishing she could go to bed, for the next morning she and Kate and Ned Mortimer and René Boissier with one or two others, were to go out for the promised ride. But duty must be done. Charades were all very well, but most of the men did not like them and one could not go on listening to Tabitha Gooch for ever. She was a neighbour of the Gibbs's who fancied herself as a diva. Besides her the other young people were also from the county, the twin sisters, Cecilia and Rachel Warrender and their brother William, a soldier on leave, and a few slightly older Maltravers girls who had, as yet, found no husbands. Kate's suggestion was amusing, Sarah thought. She had seen her friend being gazed at by Ned Mortimer and felt a twinge of unease. Kate was too clever by half for him. Her friend was a decided social success, though Sarah sensed that Kate had not gone down quite so well with the county girls who distrusted her accent and her mannerisms.

The company finally drew lots for who would hide the slipper and it fell upon Guy Demaine.

'You know the rooms, Guy – everywhere except Mama's drawing-room and the servants' kitchens and bedrooms. We shall count five hundred and then come to look for you. Whoever finds you takes the slipper and hides it somewhere else and you come back here, and so on.'

'But what do we *say*?' wailed Cecilia Warrender.

'You challenge whoever you've found: "I challenge you to produce the slipper" – if he can. Whoever it is gives it to you and comes down here. If the challenge is wrong, you yourself have to fall out and come back.'

'It sounds like a children's game,' said René. 'Is this how the aristocracy of England spend their evenings?'

'Have you anything better to suggest?' asked Kate. 'If not I shall go to bed.'

'No, no, I'm sure it will be great fun,' he answered with a shrug. 'I hope to be out first and then I can sit and smoke my cigar.'

The young people looked like children, he thought. Perhaps it was a curious English method of running away from others or having a delightful tête-à-tête in a broom cupboard.

'I cannot run,' objected Mrs Harper, one of the chaperons, a widow of some forty summers whose delightful instructions were to 'keep the young people happy while we talk business'.

'Never mind, dear Mrs Harper,' said Kate. 'It is not necessary for you to hunt the slipper. You will be far better off toasting your own by this nice fire.' The other chaperon had dozed off during Miss Gooch's last song and did not wake.

Marianne was hoping someone would soon find Guy and then she could come back to the drawing-room and engage him in conversation while the rest were absent. Thinking of this, she decided not to follow the others who had sat counting their five hundred, eyes shut as if in some strange pagan rite. She would go upstairs and fetch her album, for Guy and Ned had promised to write in it. There was a sudden scuffle as '499, 500' were finally chanted and the mob of young men and maidens each took a candle and disappeared. Kate had an idea that Guy would be hiding in the conservatories which opened out from one of the downstairs rooms by way of a little green door in the wall. She had shown it to him only yesterday, and thought he would be too lazy to investigate the first and second floors. He had looked rather romantic holding the red slipper (provided by Sarah) in a puzzled sort of way.

Ned Mortimer was at her side holding a candle before him like a marching saint with a cross. Bother, she thought.

'May we search together?' he said. 'I don't know the house too well and am afeared of coming upon the housekeeper darning socks or Sir Theodore asleep with his claret.'

They had all drunk wine during the dinner, but the fumes were long gone from Kate's head. 'You may follow me,' she said grudgingly. 'I have a feeling he may have gone among the potted plants. Have you seen the glasshouses? They are very fine.'

'I prefer the flowers of nature,' answered her cavalier. 'But we may as well.' What splendidly *dégagé* manners these Americans have, he was thinking. They opened the door to the conservatory together, she allowing her hand to stay just a fraction of a second on the knob as he came forward to open it for her.

Meanwhile, Marianne had gone up to her room and come with her album down the back stairs. Various screams and giggles wafted back from the front of the house. Had they found him? She turned the corner of the stair and saw a door before her that she had not noticed before with knotty wood and iron hasps. This part of the Hall was very old and for the most part the servants lived there. It was probably the housekeeper's room. She tried the door quietly and entered a small chamber with uncurtained windows and the moon streaming through the panes. At the same time she heard the sound of breathing. She put down her candle and advanced towards the drawn curtains. There was a male boot, she was sure, poking out from behind the drapery. Only one person had that boot, an elegant Parisian confection. And she had intended to wait till someone else found him and go and wait for him in the drawing-room! This was much more romantic. But her throat felt dry as she quavered, 'I challenge you to produce the slipper,' feeling more than a little foolish, but excited. Guy Demaine threw aside the curtain. He had snuffed out his candle and was a dark shape against the moonlight.

'You may have it,' he said. 'And now I can go and pay my respects to Mrs Harper and get out of this confounded silliness, as you English say.'

'Oh, please – don't go – ' she said, surprising herself. 'It is so rare one has the chance to talk. I brought my album, that why I was on the stairs outside – I was not really looking for you.'

'But the game will stop if they can't find me,' he grumbled.

'I think they are all very content to be chasing each other in the dark,' said Marianne.

'Ah, you make it sound like an excuse. Bring your candle up – no, I'll fetch it.' He did so and put it on a table near the window. 'Now let me look at you, Miss Marianne. What shall I write in your album? What do most people write?'

She answered, 'Some stupid compliment or a joke – it is the custom.'

'You are not really very enamoured of albums and games then?' he said after a silence. 'You are a serious girl. You ought to live in Paris – we don't need slippers and albums as excuses for little encounters in the dark.'

Oh dear, perhaps he mistook her unconventionality for flirtation. She must disabuse him. But she might not have another chance to impress him. What could she do or say? '*I* am not used to little encounters in the dark,' she said. 'I expect you are – some people go hunting in order to have conversations and others sit in conservatories with their chaperons.'

'It is not much fun being a young English lady, he said after a pause. Here's a rum one, he would have thought if he had been English.

'I shall have to go and hide for someone to find *me* now,' she said dolefully.

'Oh, we could hide together,' he said. 'There is a cupboard a little further down the stairs – but I'm afraid it would only take one fairly small person.'

'Then I shall hide there and you can go back to the drawing-room and suggest to someone where to look so I am not left for long imprisoned.' They were whispering as though they were intent upon some mutual plot.

'Ridiculous!' He echoed his thoughts. 'Am I not allowed *une gage*, what do you call it – forfeit? For having been found – or is it not you who pays it?'

'I don't think there are forfeits for this game,' she said. 'Look at the moon. Isn't it beautiful? And the park. I wish I lived here.'

'Where *do* you live?'

'Oh, I was at school with Kate and Sarah. My parents are not far from here – in the next county. We used to live in London, but Papa had some problems . . .'

'And Miss Kate? Does she live in London?'

Marianne explained and thought that Guy took a flattering interest in all she had to tell him about her friend. Finally, as they made to leave the room, he took the candle from her, put it on the floorboard and said, 'I shall kiss your hand, fair Marianne. I have enjoyed our little conversation. Perhaps we could continue it when the others are riding tomorrow – if you don't object?'

'How could I object? It is a novelty for me to have a conversation which is not concerned with balls or dances or baubles or boredoms.'

He took her hand and pressed it to his lips. Why did he do it?

Was he just being *galant*? Was he bored? As he raised his dark eyes to hers she felt a shiver go down her spine as though she had had a sudden fright.

'There,' he said, releasing her small hand and smiling. 'Run away and hide in the cupboard on the left down the stairs and I shall take your candle and send a young lady to find you.'

As they passed through the door he pressed her hand once more and she was in the seventh heaven of delight.

4

Marianne had not thought much about the differences between man and man before, but was forced to consider this when she was discovered in the stairs cupboard by Ned Mortimer, who seemed pleased to find her and rather gingerly helped her out. He had passed an entertaining quarter of an hour with Kate in the conservatory till she had decided it was time for them to join the game. He had (unwillingly) left her in the drawing-room looking at a book of pressed flowers before venturing out for a mild exploration (she said). When Guy Demaine returned from being discovered, he found her standing at the window, one hand on the fold of the long velvet curtain and with a pensive air. How different she was from little Marianne, he thought. Marianne was a pretty girl and ripe for flirtation, if one discounted the over-eager manner and rather breathless excitement she seemed to generate around herself. But Kate was more of a mystery. Her poise frightened him and attracted him in equal parts.

'Shall you join the fun?' he asked as he moved up to her with a glance at the dozing lady by the fire. The other chaperone was deep in a book.

'I don't really care to,' she answered, with an amused smile on her face. 'Who found you?' she asked idly.

'Marianne – she was not really looking for me either – was inspecting the back stairs, I believe.'

'Marianne is an attractive girl, don't you think?' she ventured.

He was rather taken aback. The American's outspokenness was more like that of a Frenchwoman.

'Certainly there is a charm about young English girls – '

'Does it put you on your mettle?' she asked, laughing.

'Oh, Miss Mésure (he pronounced it in the French way), there are charming girls too in France, you know – but they are not let out by their Mamas.' He was troubled a little as he spoke by a feeling that Marianne was more than just 'charming', but could not put his finger upon it.

'Marianne has hidden depths,' said Kate.

He looked at her with slightly narrowed eyes. What *this* girl had was a fortune, if Marianne were correct. He must play his cards right. Not that he thought Miss Mésure was interested in him. Indeed, she now seemed to dismiss him with a whisk of her fan, saying, 'I shall do my duty and hide myself. It might be fun to go behind the draperies of the curtains here, do you not think? Nobody will think of that and I shall be able to look at the moonlight in peace.'

He took this as a desire for solitude so went into the next room where brandy and soda were laid out for the gentlemen.

Next morning, when Sarah and Kate had gone out hacking with Ned and René, Guy stayed behind to sketch the house from the drive. He had the firm intention of painting a few pictures of an English country house to show in Paris. But he wandered in not long before lunch, hearing music emanating from the smaller of the drawing-rooms where someone seemed to be practising. It was Marianne, who had hoped thus to secure his interest.

Guy was feeling rather sentimental that morning and a little bored. He had been humming the tunes from last season's great success *La Vie Parisienne* and remembering the little midinette whom he had taken out last year. She'd disappeared, or was dead, or had gone off with someone else – and he had been rather sorry. England had not entirely succeeded in taking his thoughts away from Paris where he had been paying court without much success to a rich young widow. He was not content with his sketches of Lake Hall and wished that he had gone for a ride – except that he was not a very good horseman and had not wanted to appear clumsy in comparison with his friend René, who was, or the young Englishman Ned, whom he suspected might be.

English girls were strange, Guy was thinking. He could not make them out. On the one hand they were all protected and swathed up in the trammels of class and good behaviour – more than some French girls he knew, even of his own class. But on the other hand they seemed freer in their way of talking than most middle-class French girls. It was all to do with outward appearances, he supposed. The French girls knew what was what and were quite clear that marriage was the aim of their flirtations, provided their Papas agreed and the intended was a *bon parti*. Once married, they became far more open in their dealings, especially if their husbands were elderly or had been chosen with an eye to dynasty or money rather than inclination. But these English girls, in spite of all the Evangelical rubbish that René had told him was *de rigueur* in England, were fresher and more spontaneous and more – he searched for the word – and found only 'romantic'. The droll thing was that René had intimated to him that everyone in England thought the *French* were romantic, whereas they were, in reality, the hardest-headed nation on earth. What about Americans though? Kate Mesure quite baffled him with the contrast between her free and easy conversation and her rather cool manners. He thought there was more than calculation in the way she had talked, for example, to Ned Mortimer. He supposed Ned must be rich. Sarah now – *she* was a nice girl and lived in a very pleasant house he must admit, but she was too horsey, too raw-boned, too innocent-looking for him.

He paused in the doorway to the drawing-room and listened to Marianne's rendering of a Mendelssohn 'Song Without Words'. He had still not made Marianne out. She was very attractive, just as mysterious as Kate, though in a different way. It was she who had told him Kate was rich, he remembered. Guy wished for a moment it was Kate Mesure who was playing the instrument, but Marianne would do to pass half an hour with, provided those boring chaperones did not intrude. He watched her as she concentrated on her music; it reminded him a little of the way his friend Henriette had dressed a hat, rather solemn and innocent she had looked, though she had not been at all innocent. He sighed. If only he had a fortune as well as a handsome face. He was uncomfortably aware that he was not a rich man.

Marianne looked up. She had known for some moments that Guy was watching her, and that had made her play several wrong notes. She got up and did look a little startled.

'Do go on playing – do you mind if I smoke?' he said and flung himself into an armchair.

'How did your sketching go? I thought it was only young ladies who sketched,' she said, and then hoped she had not appeared rude.

'Badly. I paint for pleasure – '

'Don't you want to be a professional artist?' she asked.

'I did once – but you know, one must not have too high expectations. Things are very much changing in the painting – excuse my English – if you want to make money you need *patrons*, but if you want that people take your work seriously, that is not enough. You must change your style – all is changing.'

'Oh, it's changing in England too – don't you love the painting of the pre-Raphaelite Brotherhood!' cried Marianne. 'I don't know much about the new French painters. But you must show me *your* paintings. The park here has an atmosphere – I am sure you could catch it in paint and then make an album of English Parks to sell in Paris,' she rattled on.

He laughed. 'Ah, the English have not yet caught on to our new way of painting, but your light is just right for it, so green and rainy. Wet light – can you say that?'

'Wet light,' echoed Marianne, putting her music away and preparing for an intellectual conversation. He was thinking, English painters are so moralistic, but said instead, 'You do not ride, then?' as she rose to close the piano.

'I don't like it very much. I am hopeless. I expect Kate will be enjoying herself, and Sarah – we could play croquet this afternoon, I expect.'

'Ah, that is a very English game – and ladies can play it too.'

'Oh, English ladies, some of them, are very sporty,' she answered

' "Sporty" – what is that? – *qu'elles aiment les sports*? It is true English girls walk more than do French ones. But would you rather not go on playing – do not let me stop you!'

'I would rather talk,' she said simply. He was so handsome sitting there and she might never get another opportunity to be alone with him.

'I wish it were autumn,' she said, staying on the piano stool. 'We have the whole season to get through and it will be very exhausting and very boring.'

'Tell me – this "season" – it is like our Emperor and his Empress, their court? It glitters! All say the English are the richest nation on earth and their season is of the most brilliant – the dancing, the parties, the races, the cricket matches, the balls – '

'Our Queen is still in mourning, but society goes on.' She wanted to say, 'It is a marriage market,' but did not quite dare.

'And you, what will you do then? *After* the season – or does it go on all year? I have heard from René of the hunting and the shooting and the yachts – I have been to England only as a child – I am very "French", you know.'

'We are not yet really out,' she answered. 'Sarah was to be presented at court so Kate's guardian and my mother have asked Lady Gibbs to present us too. You have to be presented by someone who has been presented herself. Is it not like that in France?'

'Really? I don't know.' Guy forbore from saying that he was not quite of that elevated class that spent its time at court. Of course, neither was she, it appeared.

'Of course, you are a "professional gentleman".' she smiled, remembering. 'Anyway, men don't have to be presented here you see – they just accompany the ladies and enjoy themselves.'

'But not young men as young as you and your friends, surely?' he asked. 'They would still be at school or studying or entering the Army – I have the impression that the young men in your season are older than the ladies and are looking for wives. Am I not right?'

Marianne had never thought of it that way round before. 'I suppose they must be. Of course, marriage is always said to be for girls, but I suppose the men want it too, so they come to be introduced to girls younger than themselves. Naturally it is the Mamas who arrange everything – the girls' mamas, I mean. The fathers just pay.' Really, what an odd conversation and even a little indecorous. But he really seemed interested. It must appear strange to a young Frenchman that all this should be going on. Was it all for the benefit of the girls? Surely not. It was for the benefit of men. So many things were arranged specially for them, the hunting and the shooting and the sports and the sailing. She was silent, thinking it over.

'English girls are very frank, I think, but not quite so frank as Americans?' he offered.

'I do not think *all* English girls are frank. We are brought up to have nothing to say for ourselves, but it is mortally boring – we can't talk to men or even to older ladies without appearing impolite. We are supposed to have nothing whatever to say for ourselves until our first dance or dinner-party. Then we must entertain and be amusing. It must be greatly tedious for the men, I think.'

'What would you like to talk about?' he asked, turning upon her the brilliant brown eyes and the teasing look.

'Oh, books and poetry and music and painting and politics and ideas, you know. The English do not talk so much of these things. Kate says that French people do – though not perhaps young ladies . . .'

'Most young ladies don't really want to talk about politics or books, surely?' he asked. 'I have not met many who did.'

'Some do, but it is not done. You see, the only things worth talking about are the ones you must never introduce as subjects of conversation in public – '

'So you are confined to discussions with your young women friends? I should like to hear Miss Sarah on politics,' he teased.

She looked offended. 'Sarah may not be interested in that, but she knows a lot about gardens and horses – I mean like a man, not to chatter about, but because she really *knows* about them.' He noted her loyalty to her friend.

'But *you* would prefer to talk books, *n'est-ce pas*?'

'Yes, but probably you would prefer to discuss business?' she said slyly.

'Oh, business – no! It bores me. I am supposed to become a lawyer, you know, but I am not really interested in that.' He had slightly changed his story, she noticed.

'What *do* you like doing then? I mean apart from painting?' she ventured.

He looked at her with an amused expression. What would she say if he said he liked opera and dalliance and money and fashion and talking to older ladies and making love to younger ones? 'Opera,' he replied. 'I love the opera – and I enjoy giving parties – now that is something I really enjoy. *J'ai un talent pour cela*. And I love Italy and the sun, and travel and seeing new places.' He did not tell her either that he knew he was a dilettante, had not the staying power to work towards a profession. 'You ought to hear the music of the new Offenbach opera.'

'And do you read?' she asked after a silence. She had never yet seen an opera of any kind and felt cross that she could not pronounce on Offenbach.

'Well, I could not say that I have spent my youth reading our classics,' he replied. 'But I have read a good deal of Balzac – now there's an author for you – '

'And Dickens?' she interrupted. 'You have read our great Dickens.'

'A little,' he replied.

'And you must have read the poets of the Parnasse – and Mürger – '

'Mürger, *mon Dieu*, that is not a writer for little girls! I did not know anybody in England had heard of him. Now Dumas *fils*, you should see his plays – '

'I have read the exile, your Hugo,' said Mary. 'Oh, he is wonderful – do you read him? – and Monsieur de Vigny and Lamartine – '

'They taught you all this at your school?'

'No, only silly poems. But Sarah and I discovered Lamartine and Hugo from a French girl who was with us.'

'Ah – you may recite me some then,' he said and crossed his legs.

'You are teasing me,' she said. 'My French accent would not please you. Kate speaks your language very well as she has travelled a good deal in Europe. But mine was learnt only from a Mademoiselle.'

'Nevertheless – '

'I would rather play the piano – ' said Marianne.

'Then if you will continue to play the piano I shall recite you "Le Lac".'

'Oh, yes, do speak French to me, I should like that. And if you will not laugh at me I will try to speak it too.'

'Listen then, Miss Marianne.'

'Please call me Marianne – I hate that word Miss.'

'Then Mademoiselle Marianne *je commence*,' He uncrossed his legs and laid a hand on the side of the armchair. 'You may translate after each verse,' he said.

Marianne listened and his words seemed to belie the light grace with which he uttered a poem she knew well. Here was a man speaking to her of love, albeit through the words of a poet. Marianne had always imagined her ideal man conversing with her of both literature and love. Indeed she hardly distinguished between the two. It was very romantic, but unfortunately he had only spoken two verses when Sarah came in. Guy stood up and gestured her to a seat. 'We were just amusing ourselves before luncheon,' he said. 'And Mademoiselle Marianne is to speak French. She has already played some charming music.'

Sarah smiled shyly. 'Kate and Ned went on ahead,' she said. 'They are returning by the fields on the Westholme side.'

'I must go and prepare myself for luncheon,' said Marianne.

'And so must I,' said Guy. 'We shall resume our conversation later.' He bowed to them both and went out.

When he had gone, Marianne turned rather excitedly to her friend. 'I have never had a proper conversation with a young man before except with Mr Hopkins at school. Isn't he charming? and he paints and likes opera and was about to recite the whole of "Le Lac".'

'Ah, Marianne,' said Sarah. 'You are looking quite carried away! He is pleasant company, I know, though he doesn't find me so.'

The two girls went upstairs together, Marianne pondering on all he had said. Had he been flirting or was that just her imagination? She had not been flirting, had she? The chaperones or her mother would have thought she had gone quite far enough. She did hope that she had not sounded silly. She had only spoken the truth. She did hate the idea of society. She did wish she could talk about books and ideas and even politics with men. Would she ever be able to do that?

That afternoon Kate and Ned seemed quiet. They had come in rather late for lunch and Kate had refused to play croquet afterwards pleading exhaustion from riding.

She and Ned had also begun to talk of the season when they found themselves together after half an hour or so's riding in the morning. He had cantered up to her as her horse stood reined in by a hedge with Kate adjusting her hat. She had thought he would follow her; she found him restful, but rather a bore, and was determined to see how far her dominion over him would stretch.

Ned was thinking she looked so elegant, so beautiful on her brown mare. But also she frightened him a little.

Kate, however, knew how to flatter if necessary. He was an unconscious flatterer himself, though he only spoke the truth. There was no one else in sight and Ned said, 'You look as well in the saddle as I'll wager you'll look at your presentation.' Then he stopped, rather overcome.

'Yes,' Kate replied lightly. 'It would be a good idea to be presented on horseback instead of having to endure the rigours of waiting in line trying to keep your train uncreased among hundreds of nervous girls.'

He shot a look at her and reined his horse in beside hers. René would have said, 'You sound world-weary'. But Ned said simply, 'I hope it will be worth it.' They went on to talk of cricket and Henley and Goodwood and Kate tried to look interested. In fact she *was* looking forward to going to the races, especially to Ascot. It would be fun to be accompanied by Ned. She was summing him up to herself – quite nice-looking, but shy. Needs bringing out. Very English and honourable, I don't doubt.

They set off again by common consent and reached a bridle path. Through the high hedge Kate espied Sarah galloping alone. Saying 'Follow me!' Kate sped down the lane to where it met the field at the far end and reined in again. Ned came up more slowly and Kate pointed with her riding crop to Sarah, who seemed to be practising going round in circles in the field. She espied them as she came round again.

'Have you seen the Miss Warrenders? We have lost them. I came on ahead – I thought you must have gone home.' She coloured a little, looking at Ned.

'I think they turned back,' said Kate. 'I wanted to go a little further, but lost everyone but Ned.'

Ned came to stand by her, leading his horse, a fine bay that wanted to carry on with his ride.

Kate was thinking, I could make this man want to marry me – and probably marry me – but I don't want to marry him. I'd like to make him fall in love with me though. I wonder why?

Ned was back on his horse and Sarah rode away with one backward glance at Kate which said plainly, 'I will leave you two alone.' Poor Sarah – at least she had got a little time to herself, though she would be off now seeing to Cecilia and Rachel, worried they were not enjoying themselves. The other riders were nowhere to be seen and Kate was soon off again over another field. 'Why should I feel I want to tame him?' she was thinking. 'He is shy, not savage.'

Ned had been perfectly well aware that Guy Demaine had had his eye on Kate and was startled to find himself alone with her. Guy was not one of the riders. Kate would find him a bore, he thought. But he was fascinated by her. Especially when she dared him to jump the fence at the end of the next field.

'No – you must not,' he objected. 'You do not know your mount well enough.'

'You forbid me?' she answered with a glint in her eye.

'Oh, no – I could not forbid you, Miss Mesure, but I should be responsible if you broke your neck.'

She laughed, but decided not to jump. 'But I would be responsible for you breaking yours,' she said.

He looked puzzled. 'Women cannot take the responsibility for men's miscalculations,' he answered. 'And I would not jump in any case – my own mount is restive. I think we should go back.'

'Ah, well then, if you will not jump, neither will I – but let us race each other back to the park.' She was off and he followed.

Kate was rather cross, she would have liked to make him jump. Then she would have done it herself. Perhaps one day she could make him do something. Ned beat her back by a short length and they both took the horses into the stables where Kate insisted on overseeing their dressing and rubbing down. Ned Mortimer would be a hard nut to crack and would probably bore her in the end. So she concentrated on René Boissier and took him as partner in the croquet game the following afternoon, spending the evening observing the company. René observed her observing him and thought he had understood her calculations. Her present seeming lack of interest in his friend Guy was working well, for Guy was trying hard to impress her. The little Amberson girl was doing the same thing to Guy too and Ned Mortimer was talking horses to Sarah. The others had all now gone home. Soon it would be time for the whole company to disperse. René stroked his chin. It might be better to spend another day in Oxford. Young girls were rather a strain, though he liked his cousin Sarah. Both he and Kate noticed the slight tension that seemed to have invaded the party when cards were played after dinner.

'The little Marianne was making eyes at you,' said René as he and Guy undressed later in the room in the tower, a room which René found cold and inhospitable. Guy said nothing. 'Mademoiselle Mesure is a good player, is she not?' said René again.

Guy laughed then as he took a drink of the brandy thoughtfully left for them. 'She has money, but needs an introduction to continental life, I think,' he said. 'I wish I had your fortune, mon vieux.'

'Oh, there are problems with *les vignes de Papa*,' said René. 'Rich girls don't need rich husbands, do they? I wonder what they do need – '

'I expect she will marry that boring Ned,' answered Guy. 'The English are so dull – '

'Not Miss Marianne, though.'

'No, *les filles anglaises* are better than their men deserve,' yawned Guy blowing out his candle.

5

Marianne Amberson had fallen in love with Guy Demaine. What else could this overpowering yearning be, if not love? She knew he was not her sort, was probably not even as intelligent as she was. But this was an ignoble thought and she strove to counteract it by ascribing as many qualities to Guy as she did not possess herself. She did not quite expect her feelings to be reciprocated: it was enough for the time being to enjoy the fact that he existed and that she could look at him and weave fantasies around him. He was certainly handsome, and he had flirted with her. She confessed to herself that when they had talked during that silly game and when he had come into the music room, she had certainly wanted to attract him. She wanted to show him that she cared for him. Men were not, of course, supposed to become attracted to girls who showed interest in them. She ought to try to be cold and haughty. But she could not. The very sincerity which had always informed even her acting, steered her away from the idea of intrigue or from playing hot and cold. She had observed Kate doing that and it had amused but disgusted her a little. There was Kate, playing with Ned Mortimer's affections – it was obvious – and without caring for him at all. Guy had also noticed Kate's provocative manner with the Englishman and compared it with *la petite* Marianne's passionate glances at himself, but he overestimated Marianne's experience of the world just as he underestimated Kate's.

'Why is Kate so cold to Ned?' Marianne asked Sarah as they sat in Sarah's little dressing-room the day of the men's second visit to Oxford. Kate was with Sarah's aunt inspecting the gardens. She wanted to bank up a considerable amount of good will from Aunt Charlotte in order to have her own way later.

'Kate is not cold with him, she is flirting with him,' said Sarah after a pause.

Marianne's astonishment would have made Guy laugh. 'But she doesn't feel anything for him, Sarah. Does she just want him to find her attractive?'

'That is the game most women play – haven't you noticed?' replied Sarah, threading a needle,

'A funny way of showing you are attracted to someone. I can't understand that.'

'Oh, I don't know she is really *attracted* to him,' said Sarah, puckering her brow. 'She likes to feel a certain power, I think. I would imagine she is much more "attracted", as you put it, to Guy Demaine.' Sarah had finally decided this was the case.

'Sarah! I can't believe that! And yet you are usually so wise. How do you know?'

'I don't *know* – I can just feel it. If Guy marries anyone it will be for her money.'

Marianne's heart sank. It was unusual for Sarah to pronounce and so when she did Marianne was inclined to believe her. She hoped that her friend's percipience had not gone so far as to guess her own growing passion for Monsieur Demaine. She had no money. If Sarah were right, Guy would never want to marry her. Kate could have anyone with her American fortune.

'I can't see him as a fortune hunter,' she said as mildly as she could manage, trying to look as if she didn't care.

'I think Frenchmen marry even more for money than do English gentlemen. They are not said to marry for love,' replied Sarah.

'I thought that was just the *girls* of France – their arranged marriages and all.'

'What is our society but a marriage market?' asked Sarah, bending her head over her silk camisole. It was not new. Sarah's garments were not of the latest fashion or of the most expensive. The Gibbses had gold-rimmed dinner plates but a lack of liquid cash.

'Then the richest men will not mind the poorest wives and the poorest men will look for the richest. And the rich girls need not bother to look for anyone, whilst the poor – no one will want them anyway.'

It was the last conversation of any import Marianne was to have with either Kate and Sarah for some days, for May approached and all were to go to London, Kate to her uncle's, and Sarah with her mother to Aunt Charlotte Leverton's town house. Her husband was an old-established city merchant. Marianne was to go too, but must first go home to prepare. She was not sure how much money her parents were paying to Lady Gibbs for the privilege of having their daughter presented. Sarah did not mention the matter. It had all been taken out of their hands by parents. The Gibbses were pleased to help their daughter's friends and it was the custom for a little present in the form of ornaments or game or invitations to stand in lieu of money. Marianne wished her parents would not insist on presenting her, but they were keen to do their duty and hopeful that Marianne, if she could be persuaded to look a little more conventional and to talk less, might achieve a marriage that summer. They could then feel relieved of the burden. There were two younger sisters growing up as well as three boys whose school fees were a drain on Mr Amberson's finances.

Kate, of course, found the whole proceedings amusing. Lady Gibbs could offer an entrée and *she* could offer a handsome cheque to pay for the rehabilitation of the neglected gardens, so honour was satisfied all round. She had also decided to give a small party herself at her uncle's on Presentation Day. She was looking forward to the season herself, but was aware that Sarah approached it with a mixture of boredom and duty and that Marianne's sense of duty was mixed with nervousness. She knew also that it embarrassed Marianne more than a little that her parents had sent her to be finished at Madame Duplessis's principally on the off-chance that she would meet there some daughter of the aristocracy whose Mama would be willing to present her.

Her own fortune, as much as her personality, acted upon the American girl to intensify a certain hidden disdain for society's workings, and she had also conserved some democratic American notions from her upbringing. Yet she wanted to conquer it in her own way, which might not be the way of her Uncle Adolphe and Aunt Anna Chauncey.

Marianne felt she belonged elsewhere; her father's brother had become a Fellow of the new University College in London where he followed the Natural Sciences and he was the only member of her family with whom she felt any affinity. Uncle Lechmere was thought at one time to be slightly stupid until he found his vocation first of all in the Church and then, after he had lost his faith in 1859 and published a short appendage to Darwin, as an academic. Uncle Lechmere did not like society. He was a radical and a reformer and made his brother, Marianne's father, rather uneasy.

She wished she were a young man and then she would have been given a better education. She had, however, decided to make the best of her entrance into the adult world, but meeting Guy Demaine had thrown her into a dizzy turmoil of dreams and fantasies. If Guy loved her she might escape the stifling atmosphere of England and go where (Kate said) there was more fun and conversation. Marianne knew she should be worrying about finding a rich husband, but she was not. If she could get through one season and if Guy Demaine could love her a little, he might take her away to a different life, even one of relative poverty and then her cup would run over. In the meantime, for her, as for Kate and Sarah, the season was to be endured, if not enjoyed.

Kate's party at her uncle's in Westminster Square was the first event the three girls attended after their presentation in May. By then it seemed to Marianne that she had been in a peculiar variety of hell for several days. Not once had she seen Guy Demaine and she was moping. She did not like to ask Sarah where he might be, for she felt her feelings should now be hidden; she had been too open previously with her friends and dreaded becoming an object of mockery. All through the long afternoon of her presentation, Marianne had felt it would never happen; she would never have to curtsey to the Princess of Wales, never have to hold the train of her dress in that agonisingly difficult way, never have to stand in the overheated anteroom waiting and waiting until her head and her feet both ached and she was sure her face too bore signs of strain. Why did people make so much of it? She was not alone that day. There were hundreds of others all versed in protocol as though they were about to meet God or His Son and not a royal lady.

The courtiers were rather overbearing and bossy and stuffy and if it had not been for Kate and Sarah, who were with her, Sarah going first because of her rank, she would have run away screaming. Kate preserved a coldly amused look throughout, regarding the ranks of *jeunesse dorée* with a critical air. Marianne felt overcome; who was she among all this? Outside the state rooms her parents waited for their first glimpse of their eldest daughter now really out. Kate's uncle and aunt were there too, glossy and rich and transatlantically outspoken. And Sarah had smiled as she went up to kiss the royal hand, but to no one in particular. Even then

Marianne wanted to ask, where is René? Will he bring Guy to Kate's? Where are they? Are they not allowed to partake of all this flummery? But she did forget for a blank moment or two who she was, what she was doing, acted perfectly mechanically and did as she was told till it was all over and even the flunkeys were smiling and she followed her friends in their white finery back down the long rooms and corridors to the carriages. What she could not understand was how it mattered to people, to most people. It was like a marriage, she supposed, after which everything was permitted. Now the balls and parties and the race meetings and the exhibitions and the luncheons and the dinners . . .

Sarah had not smiled at the ceremony for any reason of mirth. It had been a quite involuntary smile, for she was thinking of the pony, Snowlight, that she had had as a child, and of a long trek she had once made with her and seen the sunrise over the snow. It must have been the white dresses everywhere that brought back Snowlight, for she had not thought of her for a long time. The pony had been dead and gone for years. Sarah knew that her dress was a little short – she seemed to have grown in the last few weeks – and she knew she looked a sight, whereas, she thought, Marianne was looking pale and pretty. Kate, in a gown that was not made in England, was showing under lace her dazzingly smooth olive skin.

Kate herself had been busy absorbing the scene to describe it later. 'Presented at Court' – it both mattered and did not matter. It was the final feather in her Yankee cap. Money had proved useful; the ceremony she thought absurd.

The three girls came out together into the courtyard of the Palace, only three among the crowded, crushed, wilting mass of white. There seemed an exhalation, not too unpleasant, of powder and sweat and youth that followed the maidens as they greeted the white sky of a rather stormy May. Mayblossom and excitement, and now: 'Is that all?' And the headier scent of flowers in hair and the smell of the dresses.

Afterwards they all went back to Kate's uncle's house in a shared carriage. 'There should be painters to catch the way we look now,' Marianne thought. They were to have their portraits photographed that afternoon, but by then the tea would be over, the long, sleepless night before would have begun to take its toll, and they would look no longer special, but just three ordinary young women.

Crowds still milled around the entrance to the palace courtyard. Poor people out to stare and point. If she wanted now to raise her chin and look down her nose at them she could. They almost wanted her to, she thought.

'I'm starving,' Kate said to Sarah. 'Did you enjoy it? It wasn't too bad, was it? There were one or two horrors though – did you see that poor child who had been sick and was quite *green*? I hope sal volatile did its trick for the necessary minutes!'

Sarah jumped herself awake. 'Thank goodness it's over. Come, Marianne, we need a cup of tea.'

After all the weeks with Madame Duplessis spent practising getting in and out of carriages, climbing stairs without falling over their own feet, curtseying and walking backwards in pretend long trains encumbered by veils and with hair dressed to wobble if one were not careful, the rest seemed an anticlimax. Not for long though. By the time the girls had repaired their toilettes and put fresh flowers in their hair, Kate's uncle's table awaited them and as Marianne walked into the drawing-room she glimpsed René Boissier at the end of the hall by the open front door. She stopped herself from advancing to him, but could not resist looking over her shoulder to see if Guy accompanied him.

She was behind Sarah and so took the opportunity to turn sideways and pinch her cheeks. Immediately her boredom and tiredness and staleness seemed to have disappeared in anticipation of Guy's presence – enough to set her heart racing. It would have been almost too much if he had suddenly appeared just then. The sun had even come out and the open windows leading to the narrow balcony overlooking the square seemed to invite her to the open world. There should be music, a fanfare. Instead there was an English tea waiting for them all. Her parents arrived shortly after and then Lady Gibbs in her court dress, looking rather bemused, but evidently in the place of honour. It was like a wedding, but who were the brides? Three brides, she supposed.

They stood around in the American way at first. Beyond, behind sliding doors was a further drawing-room with white linen and silver, a basket of flowers in the centre. Then some young men arrived, probably invited by Lady Gibbs who was constantly deferred to by Mr Chauncey – Uncle Adolphe as he was called by Kate – a tall, rangy pioneering sort of man. The young men – from the Army, the Navy and the Law – and some older men who had been attending the ceremonial as representatives of the United States of America, were introduced to the girls.

Kate twitched her pearls and advanced towards a group of them. Marianne found herself, flower posy in hand, listening to a young subaltern, without much idea of what he was talking about. He was red-faced and had thick, pink hands, but was otherwise pleasant.

'May I introduce you to Sarah Gibbs,' Marianne said as Sarah loomed beside her.

Through a gap between Sarah and the door she saw a figure come in, René, and as she turned back to listen to the subaltern and Sarah tried to look interested, Kate left her own group and was at her side.

'Guy will be here,' she said. 'Uncle invited both of them.' Then she was gone again.

Marianne touched her large hair-comb, wishing she could look at herself in a glass and that she could change out of her clumsy white satin, with its fullness all at the back and its awkward overskirt, into a plain dress, perhaps a velvet one. Her father had bought her a pair of drop earrings and her mother a fine fan . . . why should she think of clothes at such a moment?

She looked across at Sarah, unaccustomedly girlish in ringlets and ribbons and probably feeling the same. Only Kate was dazzling everyone. Marianne took her courage in both hands and said to another young man who had moved up, 'I am thirsty – they kept us waiting ages.' But as she spoke, her eyes were on the door and through it there came, at last, Guy Demaine, handsomer, she thought, than any of the other men. And surely his eyes were on her! He stood there for a moment without moving, and she could not move either nor take her eyes off him. Something pricked behind her eyes; she felt as though she had drunk one glass of wine after another and wished she could throw herself into his arms. But then Guy was taken by the host's arm and moved out of her sight for a moment and another man – why, it was Ned Mortimer – was beside her, talking pleasantly. She could have murdered him! She wanted either to be by herself, or alone with Guy. Other people were like a pain in her head. It must be the long afternoon's fatigue; she must take hold of herself. She swallowed and said, 'Let us talk to Sarah – it is really her party.'

Marianne felt strongly that although Kate's uncle was actually hosting this post-presentation affair, it was in fact Sarah's party.

Obviously it was one of the ways in which Kate's family was paying Sir Theodore and his lady back for presenting their niece and protegée, but Sarah, who never pushed herself forward, tended to become neglected. Even her own mother never seemed to notice this. Just then, before Ned could look round for Sarah, Kate came up again and said, 'Come, let us sit down in the place of honour and drink our English tea and I shall toast us all in it.'

The three of them found themselves sitting with Adolphe and Anna Chauncey and Ned turned to Marianne again. 'Do you know all these people?' he asked.

'No, some are Americans and the men are mostly brought by the Gibbses. Of course René and Guy are here,' she said. Marianne never blushed. The only indication that she was embarrassed or excited was a sudden pallor. 'Isn't Sarah looking wonderful?' she continued, when he said nothing but went on sipping his tea. 'Sarah!' she cried. 'Do join us.' All she wanted was for Guy to come up to her, but he was still caught in the toils of politeness. All the guests were moving to the little tables now.

'There will be three months of dances and parties and entertainments,' said Marianne. 'One should feel tired even to think about it.'

'You do not look tired,' he said.

'Oh, I felt it, after all that waiting around.'

'I'm hungry,' said Kate. 'Aunt Anna, where is the food?'

The guests were now being served by several hired flunkeys.

Marianne saw Guy across the room and he caught sight of her too and smiled. Ah, what a heavenly smile! – he was pleased to see her. If only it could be in a quiet room, devoid of other company. She must have gazed at him for a long time since Ned had to say twice, 'Will you have a piece of cake?' Kate was already nibbling it and Ned turned to talk to her.

Marianne felt suddenly happy, for Guy Demaine, looking debonair and rather suave, was moving in the direction of her table. What could she say to him? It was enough to have him close. She must sparkle. French girls always sparkled, they said. She would have liked to put her arms round him wordlessly and she trembled at the thought. Why could one never do such a thing? Could one say, 'You are looking very handsome today'? After all, he had said her eyes were beautiful. But as these thoughts went through her head Guy began to flirt with her. She was rational enough to know he would perhaps have flirted with any young lady on that day, but it thrilled her. How could she remain calm and cool?

Sarah was observing all this from a table a few yards away where she was trying to talk sensibly to René Boissier. Her colour was rather high – she did not enjoy crushes like this and was feeling sleepy from the excitement. The tea seemed only to animate her friends.

'So *your* party will be next week?' René was saying. 'We received the invitations only yesterday – it was kind of your parents to invite us to next week's dinner.'

'Mother has been very busy. The invitations should have gone out long ago, but she left it to her maid to post and she forgot. It will be only a small dinner, my ball was before.' As she was speaking, she was wondering who might pair off with whom. Guy hasn't a bean, she thought, he will make up to Kate I'm sure. Such calculations were not done in any spirit of malice – Sarah knew more than she ever said, and was aware that if Marianne's passion (for she saw it as a passion) for a penniless Roman Catholic were seen by Mr and Mrs Amberson, her lot would not be a happy one.

More tea and cakes were served, along with water ices and crushed melon. Then as people finished they began to circulate once more.

As at all parties, the conversation was of the lightest. Marianne kept rather quiet; she caught Guy's eye occasionally while trying to think of something to say to an elderly gentleman who had chosen her to talk to. He turned out to know her academic uncle and to have a high regard for him and their talk picked up as soon as he discovered that she was not a ninny. All too soon, however, the tea was over and the guests, mindful of the evening's entertainment, were off to their houses or their clubs. Marianne was whisked away by her mother to Lady Gibbs's and enjoined to rest. 'You must conserve your strength, Marianne – it will be the most exhausting day you have ever spent.' She was glad when her Mama went off leaving her in the Gibbs's safe hands.

It was no use pleading a headache either that evening or the many that followed. The social whirl, even for such girls as Marianne and her friends, who were not of the most elevated rank, was quite enough to make an invalid of a timid or easily tired young woman. And what was it all for? The point of it was to make a suitable match which was the very thing that conventions did not allow to happen. Every day Marianne's Mama, who was staying nearby, would enquire with whom she had danced, for she was not invited herself to the dances and parties her daughter attended and had to trust the various chaperones whom Lady Gibbs procured. 'We are not doing this just for pleasure,' Mrs Amberson would constantly reiterate. 'It is our duty – and yours – to see that something comes of it all.' Marianne could have screamed. Her mother was a vulgar *parvenue* – not even that, for she had never really 'arrived' herself. And even Father was caught up in it all. What else was the spending of his money for? They had their eye on a steady husband, one of the slightly older men perhaps, whom Lady Gibbs invited. As for Marianne's friend Kate . . . they thought her handsome, but not quite a lady, which amused their daughter. At each party or dance or outing she would look to see if Guy were there. If he were not, then the evening was a wasteland. If he were, she felt she could look at no one else. Was she behaving shamelessly?

Nothing much showed on the surface. Only the stolen glances which, it must be admitted, were sometimes caught and returned by him, for Marianne was pretty and did intrigue him a little. There were, of course, other lovely young girls and he was not ready to commit himself to marriage to any one of them. He was delightful, they all admitted, but 'impossible', since he had no fortune and no family. But as the season reached its peak, he and Marianne were thrown together a little more. If he wanted to see Kate, then he would see Marianne too, or Sarah. If the six of them, with René and Ned added to the company, were to talk together, nothing would offend the proprieties, for the chaperones could easily have wool pulled over their eyes by clever girls. The trouble was the clever girls were not all aware of the true feelings of their friends. Yet gradually, as the season progressed, each girl's feelings worked upon the other's and ambitions solidified.

6

Out of the whirl of the season, Sarah was to remember the opera at Covent Garden where she had gone with Marianne in the party of a girl from Madame Duplessis's who had left the year before but not managed to secure a husband. But chiefly she was to recall the Derby at Epsom. Ascot was too much of a social duty and would necessitate turning out in the fashionable clothes that were the bugbear of her life, whereas the Derby was less refined and one might even see the populace enjoying itself.

She had gone down to Epsom from Waterloo in a train with her father's party where the talk had been only of horses, and she had even ventured an opinion of her own. It had been a lovely summer day, the scene from the train windows one of pastoral beauty. Why was she forced to join in all that artificial London life? She was a country girl. But when they arrived on the racecourse, one wide, green expanse with its private stands and its three public ones and with the great crowd already like a human sea in every direction, she felt both excitement and a sense of impotence that she was only one little person in all this. What did the horses feel, she wondered? They probably caught the mood and were infected with it. There were what seemed miles of hucksters in temporary shops set up for the occasion, under canvas, and all round in every direction not only people but carriages of every type, and carriage horses. Most of the crowd was not in a private enclosure as she would be. The noise was dreadful and she covered her ears. It would be impossible, she thought, to find Kate on such a day. Marianne was not to be part of the day. She had expressed no interest in the Epsom Derby. But Marianne would have enjoyed seeing the spectacle – she ought to be encouraged to make more effort to get out of herself and her thoughts and see how the common people enjoyed themselves.

There were gypsies and minstrels and people dressed up in a

variety of disguises, booths for coconut shies or for shooting galleries, conjurors, violinists and anyone who had a talent to sell. She passed the stalls of food and sweetmeats and fruit, some set up in moveable cabs or carriages, and various carts selling drink. When they had by dint of their servants' efforts arrived in their private box which they were to share with some of Sir Theodore's friends, Sarah found the champagne already circulating and she accepted a glass absentmindedly, her eyes looking over the vast crowds, dense for the most part, but patchy on the edges where there were beggars and bootblacks, and pedlars, some of them women with small children in rags wearing incongruous handed-down hats which had once belonged to ladies like herself. She hoped they were enjoying themselves. It was not often that one saw so many of the poor gathered together. Sarah had never been in the East End; the nearest she had come to seeing real poverty was in the small town near their village. The police were busy clearing a way for the nobs, and standing round staring at people. Then a detachment advanced to clear the racecourse itself which now appeared from their stand like a green lagoon in the midst of humanity's surf. Soon the horses appeared, with their jockeys already seated, at which point Sarah heard a voice in her ear and turned to fine René Boissier by her side.

'Enjoying yourself?' he asked.

She nodded and her gaze returned to the field. A bell was rung to begin, but the horses were reined in more than once before a start could be made. Finally they were off, bunched together at first in the far distance then stringing themselves out as they came round. Sarah had her hopes pinned on Phoenix, a chestnut three-year-old, and thought she saw him in the leading group. But it was difficult to distinguish his yellow colours and she raised her field glasses and herself on tiptoe. Several horses had fallen and now there was a sudden shouting from the vast crowd which then grew louder and louder. Hats were thrown in the air and people were jumping up and down. In their stand no one jumped or shouted, but she could feel a tautness in the company. Her father was leaning forward as if in agony and she saw a horse come in followed by a yellow-suited jockey, taking second place. The hucksters were throwing up their arms and there were shouts of 'Hermit! Hermit has won!' Phoenix was second, and she turned to tell René, but he had disappeared. She had placed a private bet with her father on Phoenix both ways so would not lose, but she felt disappointed that at her first Derby her horse had not come in first. Had she really expected him to? Now suddenly the whole crowd streamed on to the course like a tidal wave, men, women and children, and she could see no more of the winning four, who must be led into the paddock through the ranks of police, again much in evidence.

'Well, Sarah,' said her father. 'You have made a guinea, I believe.'

'Where did René go? Did he bet?' she asked. The lady on her other side was moaning faintly and for a moment she wondered if she were ill. But no – it was only the excitement of losing. Losing or winning, what did it really matter? It was the race that mattered. Yet she had hardly seen the race! The crowd below had seen as much as she had. But Phoenix *was* second, and René was back, this time with Ned Mortimer, who was, it appeared, looking for Kate. 'No, I have not seen her. Isn't it extraordinary?' Sarah was pink-cheeked – her usual placidity had disappeared for a few minutes. What fun it would be to be a jockey oneself – though she would be far too heavy for such a race.

They decided to stay on in their stand for a time while a few of the other gentlemen went to fetch their winnings. Champagne was again brought round. It was like a play at the theatre, but more fun. 'I wish I could go down and mingle with the crowd,' she thought. 'Lose myself among them all – be part of it – ' But Ned was now talking to her, not having found Kate, and Ned was the nicest man she knew, so she made an effort. 'Are you to take Kate to Ascot?' she asked, and Ned flushed a little before replying, Yes, they were to be in Uncle Adolphe's party. Was she not to be of that company too? 'No, I have not been invited,' she replied.

'Then you shall come with me,' said René. 'Your father has asked me to be of his company on that day – you will enjoy that.'

Yes, she would go to Ascot. Now she knew she loved race-meetings and she would remember her first Derby. It was all over so quickly though. There were other races here of course, but now the Derby was won all that remained was to guzzle and drink and for the ladies to indulge in polite conversation away from the crowds. Still, it had been an eye-opener. She must tell Marianne about it. Where did all those people go when the day was over? She thought about the whole day as they returned later in another train and wished she could be returning elsewhere and not on a train but on a tall horse, dressed in Hermit's colours. In the evening the returning crowds filled the London streets, drunk and abandoned, even the children, but by morning had melted away.

Kate was to remember most vividly the day when Uncle Adolphe explained once more that half the money held in trust for her on attaining the age of twenty-one would be hers on marriage. He was trying to say in his easy way that if Kate had had a proposal he would like to know about it. She was quite sure Ned Mortimer would propose if she gave him half a chance, but was determined not to give him that chance. She did not wish to regret her youth for the rest of her life or to make a nice man unhappy, even if it meant coming into her fortune. She had someone else in mind as candidate for matrimony, but said nothing to her uncle yet.

Marianne remembered only Guy's face and Guy's voice. The conviction was growing on her that he loved her as much as she did him. Otherwise her favourite day was when she escaped 'official' surveillance and went with her uncle to a Dickens reading. Marianne was not sure if she really liked London; on the whole she thought not. Dickens, though, had been wonderful: that was what life was really about.

The reading had been in a large hall and was from *David Copperfield* and *The Old Curiosity Shop*, but chiefly from the latter. Mr Dickens was not a large man and looked older than she had imagined. His hair had once been curly but was plastered down with Macassar oil and he looked just like any other middle-aged or elderly man. But when he began to read you forgot his person and listened to his voice which was very expressive and rose when necessary to a crescendo, before dropping again to a whisper. She and her uncle had good seats and so she could see that sweat was pouring off him and that there were tears in his eyes when he came to the death of Little Nell. She shut her eyes and listened and wondered whether she would have read it quite like that. He was a showman – a different person from the man who had written the words in the first place. But he seemed to know them almost by heart. There was a slight Cockney tinge to his voice that pleased her. Mr Dickens was not of the élite of society, or, if he was, it was through his own efforts. The audience too was not very fashionable on the whole. There were girls there with their parents, girls even younger than herself, and many elderly, shabbily dressed ladies, and old gentlemen. But whoever they were, they were all held in a magic spell, in a web of words where stories were spun. She wished she could have written them herself. The speaker needed a glass of water now and then and she could see his arm was trembling as he poured himself one. When he had finished there was a hushed silence and then clapping and some stamping of feet. To see genius in the raw! How did *he* compare with the men and women she was forced to live among? Better by far, and greater. She did not want to move, sat staring at him, resolving to read all his books now, even the ones that people said were a falling off from the novels of his youth. The last one was said to be most peculiar and Marianne had borrowed it from the library at Lake Hall Park, but not yet had time to read it. What energy there had been in his rendering, and what restlessness, almost as though he saw the characters before him as he spoke. He was shortly to make a tour of America, she heard. He ought to spare himself. He would tire himself out. Now he sat on as the audience clapped and cheered and he seemed to droop for a time, then mopped his brow and smiled. 'He will drain himself to the dregs,' said her uncle. 'He does not look a well man, does he?' Uncle Lechmere treated her as an equal and expected rational conversation. She felt more at ease in his company than in anyone's.

'Have you read *Our Mutual Friend*?' she asked him. 'People say he is finished, that he can no longer please his public, I haven't read it yet.'

'He will please himself – and then posterity,' said Uncle Lechmere. 'I expect you would like tea and crumpets now?'

They had left the hall and she had hoped to catch another glimpse of the great man but he had been whisked away and the crowd had dispersed, disappointed. Afterwards, though, she listened to her uncle over tea saying that Dickens was not a comic novelist, nor a social reformer, but a romantic genius like Shakespeare. So she learnt another use of the word romantic, and applied herself (when she was allowed to remain in her room) to reading *Our Mutual Friend*, sorrowing over the fact that she knew no *young* men with whom she might discuss his work. Kate, perhaps, had read his *Chuzzlewit* and might have something to say about his portrayal of America, but Kate was too busy enjoying herself to have time for literary conversation.

Kate went to Ascot where her hat was much remarked upon,

and to polo at Ranelagh with Ned. All the girls went to the Royal Academy exhibition, Marianne disliking most of the pictures, and all went once or twice to Church Parade in Hyde Park on Sunday. The whole thing was a source of much amusement to Kate, the cuttings and the bowings and the parasols and the paraphernalia of wealth and beauty. She was remarked upon there too.

Clothes, money, social climbing, power over men, pleasure in dancing and flirting and being looked at, gossip and dreams – the season was that in some measure for them all, even Sarah. Sarah had no desire for anything to happen to her personally, whereas Marianne wanted always for her life to begin. Kate knew that a good deal had happened already and bided her time. Things could always be *made* to happen, she thought.

There came the evening when Marianne injudiciously mentioned Guy Demaine to her Mama and that lady took fright. She was amazed at the effect her few words of praise for wonderful Guy had on both her parents. At first they assumed he had asked for her, and when they gathered he was a Catholic and also poor and French they were extremely angry. Marianne was scandalised by their reaction but protested truthfully that he had said nothing to her. Then her Mama called her a shameless hussy for encouraging such a poor prospect and probably making herself cheap in the process. Marianne was about to rebut this with an indignant: 'I have said nothing to him,' when she bethought herself that perhaps she soon might, in which case she had better say no more on the subject. She suffered the rest of the parental homily with mounting indignation, however, and resolved to mention his name no more until the day he asked for her hand.

She told Sarah some of this as they breakfasted late one morning. It was now late July and they were both at a London house party after a dance the night before. There had been a storm at midnight so several girls had been obliged to stay over and were at present accommodated in the governess's bedroom on the top floor of a house near Grosvenor Square. Kate was not present; they were to see her at a garden party later in the day. Sarah was to go that same week to the Eton and Harrow cricket match in her father's party. Marianne felt flat and dull. What was the good of risking maternal and paternal wrath when the object of all your desires would not declare himself? Half of her mind understood her own self-deception, but the other, rasher, half could not desist from hope.

London was beginning to empty and there was a rush to get the last dinners and dances of the season over now, for in August many people would go to Cowes for the yachting; others would go abroad to take the waters or see the sights, and others still to country house parties. All three girls were to meet after the garden party at a 'crush' at the Warrenders' London house.

'I know I should not have mentioned his name,' said Marianne to Sarah. She had not done her hair and was wearing an old *peignoir* lent by the daughter of the house. ' "*A Roman Catholic*! And *French*! What fortune has he?" – I thought I would go mad or throw a chair at Mama, and when she said, "And he has dared to ask you?" I felt so stupid to say no, of course not, I was just telling her about an amusing acquaintance.'

'But he is not just that for you, is he, dear?' said Sarah delicately. 'And they will soon be returning to France.'

Marianne went on unhappily, 'When shall I see him again after this summer?'

'They will be at the Warrenders' ball tonight,' said Sarah. 'Their Mama was rather short of young sparks so asked my mother for the names of the delightful young men her girls had met at Lake Hall.'

'Oh! They will be there!' cried Marianne. 'Oh, I never knew! I am *so* glad.' Life seemed suddenly rosier.

'Kate will be there too,' said Sarah.

'Oh, Kate – she has eyes only for Ned. They were at Henley together you know, and much remarked upon, I dare say.'

'But Ned was at the cricket match without her . . . I do so wish you had come.'

'Ned does not care for *me*, I'm sure,' said Marianne, frowning.

'Oh, I am sure he does,' said Sarah quickly. 'He is always talking about you to me.'

'About me? But he is always talking to Kate!'

Sarah decided to say nothing about her views of Kate's intentions and afterwards was sorry she had not; it might have made things easier for Marianne.

'So, we have the garden party today and then the ball, and then, apart from tennis and boating parties and picnics in the country, there is nothing to keep your cousin René and Guy in England,'

Marianne said. Guy might even say something to her before he left. She daydreamed that he would tell her they would have to wait, as he was poor, or that she must soften her parents towards him. And yet there had only been the looks and the conversations and the squeeze of a hand while dancing and the feeling she had of silent communication.

Sarah wanted to tell her friend she must not hope for too much from Guy Demaine, but how could she? Marianne lived on hope. She wrote in her journal that men seemed to refuse to admit they were ever in love, sensing a trap. Yet they must have feelings similar to those of women. Men were human, after all.

Marianne did not know that men distrusted giving the name of love to mutual attraction and desire, knowing full well that if that were love they must feel it for many, many women. She had enough common sense, though, to know that you could not build a marriage on this feeling of 'love', that shared tastes and compatible temperaments were necessary. But how could she deny the strong feelings Guy aroused, and look for friendship and a suitable pleasant relation with a man who would be a 'good husband'? Should the good husband not also be he who set her heart alight? But she would like her husband to be a friend too, a younger Lechmere with whom she might talk freely and explore a world of the intellect. Society said that love and marriage went together, so she was doubly caught. If she were forced to renounce Guy, would she have to renounce her deepest desires and never marry? Would she have to accommodate herself to an ordinary marriage, make the best of it, and call love what was only the expression of the pleasant mutually supportive emotions that would hold together a family? Was she different from other women in wanting more? Usually she distrusted her own strong opinions, but in the matter of love, she felt she was right.

She was self-conscious in society, felt insecure, ridiculous – and yet at the bottom of her heart, not wrong. Yet she wanted Guy Demaine to say 'I love you' more than anything else in the world. How could he speak to her as he did in such an exciting way if he did not love her? *She* would have said the words. Did they mean something different if a man said them? She loved him. In her journal she wrote the truth as she saw it: she knew what she felt. To others though, she knew it would seem either shocking, or that she was a fool. And at the same time society spoke to her and said that woman's real satisfaction was in service to others, in bearing children and in being a good wife, supporting a husband whatever he did. Faithful unto death. If it were Guy she had to be faithful to, that would be exquisite, but Guy might not be the sort of man who would want it. She knew that Ned, on the other hand, though she had denied it to Sarah, both liked her and found her attractive. Would he call that love? Would society then be satisfied? No, she must have the courage to be unconventional, to be the prime mover in love, to convince Guy that she loved him, hoping this would give him the courage to match her love with his.

She did not like the self that other men wanted her to be. She was impetuous, so she must be brave enough to face Guy with her feelings. She was not like Kate who did what she wanted and hid whatever her real self was. Kate did not seem to have romantic feelings; Kate was not afraid of anyone. And Sarah too seemed to keep her feelings to herself.

From an early age Marianne had been told to repress feeling; women must protect themselves or they would be hurt, or shunned, cast out, beyond the pale, fallen. Were they right, these guardians of morality? Was it because unrepressed women were a danger to themselves and to the future of society? She wished sometimes that she did not find it so easy to give herself over to her feelings, her raptures and enthusiasms. Perhaps Guy did not like that in a woman, yet it had been Guy himself who had shown initial interest in her. She tried to recall the progress of their acquaintance. He had, she decided, rather encouraged her to fall in love with him. She need not have become infatuated, true, but it would have been heartless to ignore him – as so many women might have ignored him – and then trap him in the end. Why should a woman have to flirt a little to arouse interest, but remain cold underneath? How could one do this when underneath was a burning fire of passion? If he had known her better he might not have flirted with her. Poor Guy, he had not known what he was letting himself in for.

She often found herself thinking such thoughts when she looked at herself in her glass. She was not beautiful – quite pretty possibly – but what had looks really to do with it all? A little worm of doubt would say to her, 'If you had not been pretty he would not have flirted with you.' I wish I had been ugly, she would sometimes think. Then there need not be any deception. She could attract through her mind, her personality, not her two eyes and full mouth.

As she prepared herself for the garden party, she thought, I would like to love people the way you are supposed to love God. But God – in whom I don't think I believe – does not usually answer prayers.

She tugged at her ribbons and hoped her skirt was straight at the back. How boring clothes were. She agreed with Sarah in this matter but Sarah did not seem to worry about them, whereas *she* . . . She went off into a daydream about becoming a slight eccentric: an old cousin would leave her enough money to go abroad and live in Italy where she could stay unmarried, get to know interesting expatriates. She saw herself in a velvet beret, a waistcoat and neat boots, red ribbons in her hair, floating scarves and a mauve parasol. How silly – she had just been thinking she could not be bothered with clothes and here she was seeing herself as a romantic heroine dressed in order to attract men to her *mind*! What if once she was abroad a fearful ennui possessed her like Emma Bovary's? Kate had brought the book back from Paris and they had put a brown paper cover on it and printed 'The Sermons of Abbé Courvoisier' on the front. She smiled. Why shouldn't girls read Flaubert? Was she like Emma Bovary, a dupe to romantic imagination? Perhaps she was a little more intelligent than Emma, but that did not count if you suffered the pangs of passion. Which did she want most – marriage to Guy or freedom to be herself? She wanted to understand people, not have power over them. Kate flirted for power, and wanted power over others. She did not – she wanted power over herself and all her conflicting feelings.

Sarah was, at the same time as her friend, also considering her own heart and wondering what she would do if any young man asked her before the season ended. Of course, there was no rule that a man must propose during the season, but that was what most of them did. She had always imagined that if anyone were to propose to her they would ask her father first as was proper and she would then have to wriggle out of it. She had met only one man whom she liked more than her cousin René, and no one who seemed interested in her. That did not worry her; it was rather a relief.

7

The garden party was one of the turning points in the girls' lives, and perhaps the last time that they were all three together in true harmony. It was held in the large garden of a house in a square near South Audley Street, an unusual venue for a garden party, Sarah had thought at first. But when she arrived in the carriage with Marianne and they were shown into the house, then through a long hall to an open double door at the back leading to a long garden with a cedar tree and a lawn mown to the consistency and feel of green velvet, she thought it lovely. This was much more fun than overheated evening parties or formal balls and she blessed the host and hostess – Americans like Kate – who had rented the house for the season. The daughter of the house was one Ermine, a tall, dark young woman with an abrupt manner. Already the garden was filled with people and there were little tables and chairs scattered around where the older folk were seated. Groups of younger women wandered or stood in their silks, carrying parasols like bright coloured butterflies. Sarah glimpsed Kate first at the bottom of the garden talking to René Boissier, who had wormed his way there as usual. Kate was wearing afternoon dress, but it looked somehow too smart for a garden party. Her hair was dressed high at the back with long ringlets coming over one shoulder from behind and a camellia on the other side. Round her olive neck was a black velvet ribbon with a large cameo in its centre and the camellia was repeated on the shoulders of her gauzy white tulle gown. The square neck and the high waistline made Kate look taller. She was not, even so, in high fashion, for she was not wearing her crinoline, but a lacy overskirt. Sarah felt that her own and Marianne's dresses of muslin, which they had thought appropriate for the afternoon, made them look like schoolgirls. Marianne had threaded many ribbons in her dress and her hair and was actually looking very pretty. The scarlet ribbons went well with her blue eyes and brown locks which, unlike Kate, she had in a chignon with the same scarlet ribbons threaded through it. Sarah, in pink, felt wretchedly tall and unfashionable, but René did not seem to notice for he came up to the two of them and said, 'The fresh young English ladies – may I fetch them an ice?'

Guy was nowhere to be seen for the present, so Marianne waited with Sarah and talked to Kate. 'I love your earrings,' she said to her.

'I should have preferred a pair of cameos to match,' replied Kate, 'but there, one cannot have everything.'

'Really, why not?' said a voice, and there was Guy, dressed up to the nines in a new suit of pale grey with a rose in his buttonhole. Marianne felt faint. How beautiful he was! 'Tell me, Marianne,' said that gentleman, turning towards her. 'What is one supposed to do at a garden party?' The way he said it made them all laugh and when René returned bearing a tray of white currant ices dressed with little angelica 'ferns', they all sat at one of the little tables with a Mrs Morris, who was the girls' chaperon for the day but who had unfortunately arrived late. However, as long as she was plied with food she did not seem to notice very much and continuously fanned herself without making any tedious conversation.

Guy seemed to be rather cross with Kate for he studiously ignored her remarks and addressed his to Marianne. Ned was soon among them and Marianne sensed some irritation in Kate's manner to him. He stole glances now and then in her own direction, but Kate seemed to cast her eyes down whenever he spoke to her.

'Are you going to the Warrenders' tonight?' asked Guy, and Marianne was so pleased that he had asked, and asked her, that for a moment she could not reply.

'We shall all be there when this is finished,' she said finally, gesturing to the garden and the standing and seated figures. Was it all real? They looked like figures on some antique frieze moving and turning and stopping according to some ancient rite. Teacups were lifted and there was the sound of laughter and the murmur of conversation among the tinkling china. 'It will be the last ball,' she added. 'Of course we shall all be there.'

'Are you sad it will all be over soon, *ta saison*?' he asked.

'I suppose so. I hope *you* have enjoyed it?' she asked slyly. There was something in his manner this afternoon she could not quite place – some hidden annoyance, but not with her she was sure.

Then other young men and maidens came up and everything went on oiled wheels with, after an hour or so, the thanks, the departures, the promises to call. She was explaining the system of calling cards to Guy when Ned finally was at her elbow; Kate had disappeared.

'How many people do you think they fitted into this garden?' he said in a puzzled voice. 'No time to see the plants either – do you like flowers? Kate had one in her hair. Let me guess – was it a magnolia?'

'No, it was a camellia,' answered Marianne. 'Are *you* a gardener?'

'I am when I am allowed,' he said. 'All Englishmen are gardeners, I think, once they reach the age of thirty and live in the country.'

'I don't really like town gardens,' she said. 'This is all very fine but it's artificial, isn't it?'

'You prefer the natural – woodland and fields – we have fields of daffodils at home in spring and our cottagers have gardens with as many as a hundred varieties of old English flowers all crammed together.'

She ought to be having this conversation with Guy, she thought, but Guy would care more for parterres and French formal elegance. 'Where is Kate?' she asked. 'She seems to have disappeared and Sarah and I are going back to Lady Gibbs's sister to rest and then prepare for tonight.'

He seemed to blush, she thought. 'Over there,' he said, 'with some of her own countrymen.' They walked slowly back through the house together and joined the queue of leavers and thankers. Sarah was waiting for her with Mrs Morris in attendance, and the carriages were drawn up in a long curving line round the square. It was hot; Marianne felt it was a strange atmosphere as though thunder were on the way and yet there was a timelessness about the afternoon as if the houses had seen parties and love affairs for so many years and they would stretch forward to the future too, to next year when some other tenants would once more take the house for the season. She felt sad. Next year there would be several hundred new girls who were at present languishing in schoolrooms and then the year after, and after that.

Guy seemed to have gone but when Ned had said his goodbyes

she found him a few moments later waiting in the hall with Kate. René had joined him, and Sarah. René kissed Kate's hand as he said '*Au Revoir*' and Guy looked cross, but followed suit. 'Till tonight, Marianne,' he said. He did not kiss *her* hand. But of course, hand-kissing was rather a tribute to formality. Perhaps he did not feel formal with her. If only he too liked cottage gardens and daffodils. But one could not have everything.

'Do you think Ned will propose to Kate tonight?' asked Sarah as they dressed some hours later.

'Ned! She would not accept him,' said Marianne.

'No, I don't suppose so. Well *I* am not going to receive a proposal, that is certain.'

Had she not said something similar before? Marianne could not remember. All the girls had now met scores of young men, but only the three – Ned and Guy and René – had danced attendance more than a few times upon any of them. 'I'm surprised Kate hasn't been asked before now,' she said. 'There were so many men admiring her last week but she doesn't seem interested.'

'Kate doesn't function like that,' said Sarah. 'She will have the man she wants I'm sure.'

'But whom does she want? You said before that she was only flirting with Ned, yet I saw him looking at her rather strangely this very afternoon. He is nice, Ned.'

'Too nice, perhaps,' said Sarah. 'He obviously admires her looks and her charm. She could get anyone – and so could you, Marianne – are you still determined to have Guy?'

The direct question rather took her aback. Sarah sat on the edge of her bed regarding her solemnly. 'Yes, I think he is divine, but Mama and Papa are the obstacles, as I told you.'

'*Has* he spoken to you in any way about the future?'

'No, we have just enjoyed each other's company. Mama was only cross because she thought he *might* propose. Of course I said nothing about my real feelings,' she replied disingenuously.

'But you still have hopes?'

'I would marry him if he asked me,' replied Marianne simply. 'I think he is so attractive, so sophisticated. I would wait if necessary till I was twenty-one.'

Sarah was silent, looking for tactful words, Marianne felt. 'I am not counting on anything,' she said, to lull Sarah's suspicions, 'but

I can't bear anyone else when he is around – even Ned who is nice and who ought to marry you.'

A deep blush began on Sarah's fair neck, and spread up on to her cheeks. 'You must not say that – he has no particular liking for me, I'm sure. Men like their opposites, *he* likes girls who talk and joke or young women who are mysterious.'

'I'm not mysterious,' said Marianne. 'So he wouldn't want *me*. Do you still think Kate has designs on Guy? She seems awfully short with him?'

'She is short with Ned, too; only with René is she herself.'

'So you think she should marry *him*?'

'No, they are too alike.'

'How difficult everything is. If only we could all be free and above board and say what we feel.'

'*You* do, and it will hurt you in the end, dear.'

'Are you warning me against Guy, Sarah?'

'No, my dear, but he is not good enough for you.' Sarah looked away. 'Now be angry with me, but it is the truth.'

'I'm not angry with you. I felt this afternoon that it was all a game, that the house and the garden and all the houses and gardens we've been in and all the balls we've been to were just stage props with the same show going on every year and we were caught in it all – '

'Doesn't that then make you reflect?'

'It makes me feel there is a sort of fatalism about everything and that I shall be borne away on the wings of chance – not in charge of myself. Do you think the men feel it too?'

'No, I shouldn't think so. After all, they need not marry young to please their Mamas.'

'It is all so unfair and silly. I wish I were away from London sometimes, but then – '

'What?'

Marianne had been going to say that her passion for Guy Demaine, for she saw it as one, had the power to make her act rashly, but she could not say that to her friend who had already looked worried over her words.

'It is all to do with money,' she went on, changing tack. 'If I were rich I know I could have whoever I wanted.'

'Society is built on money and property, all this is about that – ' Sarah gestured towards their dresses already laid out on another bed in the room and to the dressing-tables where a profusion of ribbons and flowers in glasses of water awaited.

'Sarah – I did not know you were a Radical!'

'I am not, but I have always known that marriage is about property and having children to keep it all going. It is nothing to do with *your* sort of love.'

'Or Kate's?'

'Oh, Kate does not belong to it, she will find someone biddable and will do what she wants.'

'Money again,' said Marianne. 'Do you remember at Madame Duplessis's before the concert, when we were talking – it seems years ago – you said you wished you were a poor girl?'

'Then I was being romantic, my dear.'

'Well, *I* wish I were rich, indeed I do – '

Sarah said nothing but got up and brought the mirror to her friend. 'Your face is your fortune,' she said. 'Thank goodness mine is not – I shall probably marry a widower or a clergyman.'

'Don't you dream of love?' asked Marianne impulsively.

'Not exactly. As Papa has always said: "Do your duty and the rest will follow." I am resigned, as long as I can have my horses and time to do other things – '

'You deserve the best man in the world!' cried Marianne.

'And you deserve a man who might wish to understand you,' smiled Sarah.

Marianne knew she was not speaking of Guy Demaine. No matter. Guy, she felt sure, was her destiny.

There were to be dinners before the Warrenders' ball which would take place at another hired house. Kate was to dine at her aunt's as an American cousin had arrived on his way to Paris who might not see her if she did not stay in to dine that evening. Kate was in the comfortable, damask-covered, warm bedroom she had to herself and was dressing for dinner. Afterwards, when she had done her family duty, she was to be taken by her uncle to the Warrenders'. She was thinking over the events of the afternoon. Such an innocuous event, a party in the garden of a London square with her friends around, but she had felt that something was being decided. It had not escaped her that Ned had moved away to talk with Marianne. She knew why. She had refused to encourage Ned Mortimer in making her a proposal of marriage. It was not the moment for that and she did not think it ever would be. But it was satisfying to know he wanted her. She was under no illusion as to the type of wanting. Not for nothing had she spent time in Florence when relatives had left her to servants whom she had bribed when she wanted to go out. There had been *inamorati* aplenty then and she had been able to take advantage of them a little. Her lovers never conquered her common sense, though, and she had always escaped their too physical emotions. Now she was older, and wanted experience, but wanted too a marriage that would enable her to be her own woman. Ned Mortimer would be a fond and clinging husband – he might even turn into a tyrant if he were given his head – and at the idea of being under the thumb or the moral empire of such a man, her spirit rebelled. She looked forward to a planned meeting on horseback with him when they were staying at Lake Hall Park again and then she would see what happened. He would not want marriage if he knew her better. She thought she knew her own mind, but René was the man she *ought* to marry. It would make sense to marry a cynical Frenchman, older than herself, sophisticated, a man who would enjoy showing his wife off to others. Why did she not want to then? He was not an unkind man – Sarah always said he had been nice to her as a child. He was not bad-looking either, though not as handsome as Guy. Why not then? She meditated as she stared at herself in the glass, her comb arrested in the act of parting her sleek, dark hair. She put the comb down and gave herself up to more thinking. René was a man who would choose a wife when he was ready, a reasonable malleable girl – in any case he would have to be the initiator. He would marry because it was the thing to do. Nothing to do with passion. Frenchmen like him reserved passion for their *petites amies*. And René had often mentioned his Mama to her when they were making idle conversation, giving the impression of a strong-minded woman who had not entirely adapted herself to French ways. Italians and Frenchmen often did as their mothers told them, she had noticed. He might make a wife happy for a time, but he would be sure to be unfaithful. Not that that would matter in the long run though it would not be good to be seen as a victim. He would choose a steady girl as his wife, one who would be loyal and do as she was told, a woman who would leave certain matters to him while undertaking her share of the bargain. Yes, with Monsieur René Boissier the bargain would be clearly stated, if not in so many words. He would have to find a healthy girl who would bear his children, a girl who was morally irreproachable, a girl who would be content to leave him alone, in exchange for a decent household which she would run efficiently. Although half English, Monsieur Boissier was clearly not a romantic. She also found him, under his pleasant manners, rather cold. It was odd what attracted different men and women to each other. René, she suspected, had been a little spoilt as a child and would require a certain amount of mothering. He would have his own circle in Paris into which his wife would have to fit. There would be no question of her ruling him. He would expect to lead.

Not René, then? No, René might appreciate her as a wife, but it would take several years for him to realise it. She acknowledged he would be a good match for her, but life was not a cricket match or even a croquet competition. She wanted a husband who would look up to her, let her do what she wanted, but who would appear glamorous and dashing and popular in the eyes of the world. She did not need a husband to earn money for her; she needed a moral ascendancy as she already had a financial one. She pondered again, took up her hairbrush and then laid it down carefully. She must think. There could be disadvantages in getting Guy Demaine to marry her and she tried to see them as clearly as possible.

First, he liked women and women liked him. But was he really a sensual man? She thought not. He was *l'homme moyen sensuel*, but he did not have the sensual intensity she had glimpsed in René. Marianne was a fool, pitching her heart on Guy, who did not want it and whose physical charms had seduced her into thinking she loved him. Then, Guy adored parties and arranging pleasures for people – he had told them that. He loved opera and Italy, Marianne had reported; he had a nose for the socially acceptable. Had he ever loved anyone? She thought not. Did she want to catch him? Would it be worth it? Well, I do not need a protector, she thought. I have flirted with Ned and know my power over him, but he would want to protect me. I can produce the same effect upon Guy and he will not run away. Perhaps we are equals, Guy and I – equal in our enjoyment of life and in our determination to have everything upon our own terms . . . Yet it was Ned who was about to ask me to marry him, or to manoeuvre towards it. He thinks I need to marry! I had better be quick. I'm sure Guy has not told Marianne he loves her or anything of that kind, so I should not be taking him away from her, only taking away her illusions of him. She needs a bit of a shock to make her wake up and live in the real world. It would be a challenge and I like challenges. I should not be doing anything wrong – he is already attracted to me, I know. Yes, I shall marry Guy Demaine.

Her mind made up, she pressed a little rice powder on her straight, Grecian nose, bit her rather wide American lips to make the blood rush in and smoothed her eyebrows with wetted fingertips. She would try cold distance first, and slowly – perhaps it would take a month or two – let Guy realise he might be allowed to propose. But she must first ascertain that he would accept, if offered, a financial contract on the American model. He must not be offered a fortune on a plate. They would make a good couple. Ned should marry Sarah, she thought as she went down for dinner. That would be entirely suitable and surely could be managed.

Nothing happened unless you made it happen. In the meantime, she would continue her campaign at the Warrenders' ball.

8

Kate had therefore set out on the evening of the Warrenders' dance to make Guy Demaine jealous of Ned Mortimer by appearing extremely agreeable to the latter. Guy would probably turn, for a time, to Marianne for solace. That goose would talk to him about novels or evolution! Guy would eventually be ready to listen to her proposition, having expended his charm on Marianne to no avail. Her blue-stocking ways would not appeal to him. Kate had decided what she wanted, and whom, but could not resist a testing of her powers over Ned Mortimer. These Englishmen, so polite and gentle on the surface had hidden depths and it would be fun to stir them up. She was already the centre of an admiring circle of beaux when Marianne and Sarah arrived with Lady Gibbs. Tonight would be her last opportunity before retiring to the country for August.

It all went as Kate had planned. She danced with Ned whom she could see was utterly besotted with her. He had that unfocused look in his eyes that betokened desire. The only difficulty was to hold him off long enough as he was the sort of man who, like Marianne, would think a physical passion was love and ally it to ideas of marriage, as he had been brought up to do. So she shocked him, deliberately spoke of her conquests on the Continent, indicated that it was a bore having a fortune – one had to be so careful to avoid younger sons – and left the poor man in agony. But he had received the message, she thought. How long could she bait him? Guy looked over once or twice to her and when he saw she was in deep conversation with Ned, turned away and listened to Marianne, with whom he danced twice. Marianne's eyes were bright and glittering that night, but her body was soft and pliable. She was well brought up and so unlike the French girls he had made love to and he almost warned her against himself, for he did not want to break her heart, something he had formerly been accused of by girls of a different class from Miss Amberson. Surely she must know what she was doing? He decided she must, that all this talk of soul and all this evidence of passion was her way of making a conquest. Still, he was a little uneasy. Girls should not drop like ripe plums into your lap.

'I shall see you at Lake Hall Park before we return to Paris,' he said as they stood together after a particularly languorous valse. Did she expect him to propose to her or something? He took her hand – they were hidden behind a velvet curtain at the balcony end of the ballroom, a long drawing-room with a parquet floor. Her hand was slim and brown and warm and he raised it to his lips. Marianne shivered, and clasped his hand when he had released hers in a sudden gesture of abandonment.

'Yes, I shall be there, away from all this artificiality. There will be picnics and boating – it will be wonderful.' She meant wonderful to see him every day. He looked lingeringly at her with his dark brown eyes and thought: *Elle est vraiment cocotte celle-ci.*

'You must return to your chaperon,' he said lightly. 'It is not done to be seen together too much – even I understand that – but I shall look forward to being with you in the country.'

Marianne's cup was full. She stammered, 'Oh, yes, yes – ' wishing she could throw herself in his arms and be borne away. 'I don't care for the *convenances*,' she added breathlessly.

'*Ni moi non plus*,' he said and looked at her again and bowed and was gone. She spent the rest of the evening in a daze, searching only for a sight of him whoever he was dancing with, and there Sarah found her, sitting with Mrs Morris.

'You are not dancing any more?'

'I don't want to,' and the unspoken words were 'unless with Guy'.

'He danced with you twice – I saw . . . Did he . . . say anything?'

'Only that, like me, he does not care for the *convenances*,' said Marianne.

Then he did not propose, thought Sarah. Marianne was playing with fire.

Kate took good care that she refused Ned Mortimer many dances. 'See, my programme is full,' she teased him. 'But I will dance the very last valse with you.' He had to be content with that. When it came round she said, 'You see, it would be most improper to dance every dance with the same cavalier. The gossips and my uncle have their eye on us.'

Ned swallowed. 'I wish it were not just a dance,' he said boldly. 'I would rather talk to you – I – '

She interrupted him swiftly. 'You may sit and drink coffee with me till you are bored. Remember we shall see each other at the Gibbs's – unless you are to return north?'

'Promise me you will not avoid me then,' he said, feeling foolish.

'But why should I avoid you, dear Mr Mortimer – we have much in common I think. We can go riding together again, where no one can spy on us.'

He looked a little startled, but pressed her arm. And then other men came up and he saw how she was fêted and how she could marry any one of these popinjays, and went home disgruntled. He should have made himself clear, but something in her manner put him off while at the same time inflaming his animal spirits.

After that events moved quickly, but in unexpected and not wholly welcome directions. The season was over. No one had had a proposal and Marianne's parents were peeved. Sir Theodore and his Lady went back thankfully to the country and the London house was shut up when the aunt and uncle, John and Charlotte Leverton, came to Lake Hall Park. Kate's uncle stayed on in London where he had business. He was glad that Kate was invited once more to the Park, since it dispensed with any worry that she should be in London, the prey of society fortune-hunters. Afterwards, in September, they would arrange to go to Paris if she were still on his hands.

'It will only be a family party,' said Kate. 'Aunt Carrie would be bored – besides, I need a rest and a change. It has all been very interesting, but I am rather tired of London.'

'I am tired of London,' said Marianne to her parents.

They too both looked and were tired. Joseph Amberson was continually travelling up and down the length and breadth of the railway network, stopping wherever he had contacts with whom he might initiate business. Not unnaturally he suffered from dyspepsia from long hours without food or too much food at once. He could no longer afford to pay too many clerks and in any case now trusted no one but himself to do business. His wife had just had a miscarriage, but Marianne did not know this. She was, however, sorry for her mother who looked so drained and grey. Her Mama was determined that she make a good marriage and Marianne did not understand why anyone so manifestly unhappy should wish her daughter in the same state.

'Will Mr Mortimer be at the Park?' asked her mother.

'Ned Mortimer? Oh, I expect so,' replied Marianne. 'But they will mostly be older people, you know – the Gibbses do not entertain a great deal. It is just for Sarah that we are invited again.' She thought she had dispelled their fears about Monsieur Demaine and was determinedly casual about him, managed not to look in any way interested if his name came up, but talked instead of Ned Mortimer, thinking thereby to put them off the scent.

'It is a pity we cannot afford Cowes or Scotland for you,' said her father.

'Why, Papa, I am not interested in grouse-shooting or yachting! I shall occupy myself in the library at Sarah's. Uncle Lechmere has given me good advice as to my reading and I shall return home with many plans, I'm sure.'

Her father said nothing. His little Marianne was going to be a problem. There was Eliza waiting for the next season and he wished he could throw the whole idea up of getting the girls through another London summer. He had enough to do worrying about his investments and starting a new line of improved aniline dyes in his Midlands factory.

Sarah was looking forward to a calm winter's hunting, visiting the cottagers and helping her father with the accounts of the estate. She was thankful the season was over and that she could be an unmarried daughter once more, though she knew her mother was disappointed for her and that her protestations of gratitude for the three months they had devoted themselves to the cause of her marriage were not enough to make up for a lack of offers.

'There is always Mr King,' said her mother tactlessly at breakfast the day before the friends were to arrive, 'but I do not think we have room for him.'

Mr King was a neighbour, a man of forty, who farmed nearby and whose devotion to his mother was a byword in the district.

'And Captain Taylor,' replied Sarah, keeping her face straight, knowing full well the implications of her mother's remarks. She was a failure, not that they held it against her, but they would be so relieved if she found a husband. Her parents were to visit

Scotland themselves for a week or two and Aunt and Uncle Leverton were to supervise the young people. Sir Theodore insisted on his salmon fishing and since he had paid for his daughter's season, it was his turn to enjoy himself. The old housekeeper and all the servants would see to things.

Her brothers were to accompany their parents to Perth and she looked forward to entertaining her girlfriends. The Frenchmen were to stay only for a few days this time. Ned Mortimer had accepted for a week or two before he too went off back to the North of England. Soon everything would take on its accustomed rhythm, and autumn would see René and Guy safely back in Paris and Ned in Westmorland, thought Sarah, being a little uneasy (though she did not betray it to her mother) about the undercurrents that were swirling round Marianne and Kate. It could all be cleared up once they were all easy again together, almost like brothers and sisters, she felt, with none of that hectic gaiety she so disliked in town.

It was a cloudless August and the wheat sheaves were already standing in the home fields waiting to be carted away to the barn. The park was covered every morning in a light mist, the trees heavy and green and the sun shining by eight o'clock with the promise of another perfect day. They had already picnicked once down by the stream and gone for long walks during the first week. Marianne and Kate had arrived together on the Monday and there had been only the family to welcome them before Sir Theodore and his lady went away the next day, leaving Uncle and Aunt Leverton in charge. The other visitors were to arrive on the Wednesday. Ned came first, riding over from Oxford, polite as ever. Sarah's father had hoped to see him before he left for Scotland, but consoled himself with the thought that perhaps now in her accustomed surroundings his Sarah might receive encouragement. He fully expected a letter from the young man at the end of two weeks asking for her hand.

The next day the train brought Guy and René to Oxford and they were conveyed in the second-best carriage to Lake Hall Park. They were silent on the journey, both aware of their shrinking pockets and the necessity of doing something about it. But what!

At dinner that night, served in the small dining-room and presided over by Sarah's Uncle John, the talk was desultory. Everyone seemed held in a waiting silence and Sarah was reduced to making general conversation, a practice which she did not find easy and which usually fell to her mother. Aunt Charlotte was not much help either for she was so preoccupied with seeing that the *couverts* were adequate and the servants organised in the absence of their mistress that she was a little distracted. The stay at the Hall, a return for their lending the family the London house, was a change, but too much of a responsibility.

Sarah's attempts at conversation were noticed by René and applauded silently. The devil take Guy – he was being uncommonly morose. There was the little Amberson girl casting her big eyes at him and all he could think about was Miss Kate Mesure's fortune. In fairness Miss Mesure's fortune had also been considered by René himself, but he judged it not worth a lifetime's servitude. He thought the girl rather cold. Ned Mortimer was quiet too that evening, only occasionally uttering some commonplace, but glancing at Kate from time to time.

There were twelve of them at dinner – the three girls and the three men, Uncle and Aunt Leverton, a couple from the next property in a valley that wound round the park and two spinster ladies whom Lady Gibbs had begged her sister-in-law to entertain when she was temporary châtelaine of Lake Hall. The servants had made an effort so that their master and mistress should receive a good report from Mrs Leverton, but as the courses followed one upon the other, the company seemed to grow more silent. Kate seemed unaware of Ned's glances and applied herself to picking daintily at the plentiful food. Only Marianne offered more than platitudes, but even she seemed preoccupied. They were all glad when the white soup and the roast beef, the boiled chickens, the curried rabbit and the stewed wood pigeons were sent away along with the boiled potatoes and stewed celery, and the pudding course arrived with its plum pies and apricot jam tarts. Kate was thinking how perfectly dreadful English food was, how quantity rather than quality seemed the rule, even in a good establishment, and the Frenchmen, though they had tucked in with the appetites of masculine youth, were thinking that a few sauces would not have come amiss. At least there was plentiful claret, though the women took watered white wine. At last the girls and Aunt Charlotte rose with the two spinster ladies (who had feasted well), and the gentlemen were left over their port. In the drawing-room Kate was restless, prowling from window to window.

'Would you not like a walk, Sarah?' she asked her friend.

Sarah was surprised; walking was not exactly Kate's favourite occupation.

'Would you come too, Marianne?' But Marianne said she was going up to her room with a book and please to excuse her to the rest of the company, but she had rather a headache and if she were to enjoy tomorrow's picnic she had better rest. It was an excuse. She dreaded spending another evening playing some insipid card game.

'If you have a headache it would be better to get some fresh air,' said Sarah.

'Oh, well – for a few minutes then. What can those men have to talk about?'

'The usual things – at least I don't know what René and Guy talk about when they are together. I suspect they will have to listen to my uncle telling a hunting story. Ned will help them out. He always knows what to say.'

'Ned is kind, I think,' said Marianne as they walked along the terrace. It was not yet quite dark.

'Yes, he is practical,' answered Sarah, thinking, 'And would make someone a good husband, but it will not be me.'

'Where shall we picnic tomorrow?' asked Kate, following her own line of thought.

'We can take the fly along the drive and then walk through the woods. The pony can carry the picnic baskets and be taken back by the groom,' said Sarah who had thought it all out.

'It will be a formal picnic then – you do not take the servants?'

'I thought it would be better to do without a fuss – we three can go with the men. René has shown me his straw sailor hat which he bought in London for such occasions,' Sarah answered, giggling.

'If there are going to be wasps I shall wear a veil,' said Kate.

'Some more of uncle's friends are arriving tomorrow, so we shall be well out of the way.'

'It is kind of your parents to let them entertain here,' remarked Marianne.

'I bought a new hat,' said Kate. 'I must show you. Girls, soon hats will be so small one might as well wear an egg-cosy.' She shook her loose back hair which she wore *à l'américaine*, and tossed her head.

'Don't you ever get bored in the country, Sarah?' asked Marianne. 'I mean,' she went on, thinking she had been rude, 'I would like to read and spend my day in your library, but I suppose if you ride and walk your days are healthier.'

'The day after tomorrow I thought we might go out for a ride,' said Sarah. 'You can stay in and read. Of course, it depends on the weather, but it seems the warm days will hold for a time.'

Neither of her friends was uttering what was really preoccupying them. When could each find the time to be alone with the man of her choice far from the prying eyes of uncles, aunts, chaperons and neighbours? The two Frenchmen would soon be gone, and Ned too, and the rest of the year seemed to stretch on ahead in one wearying succession of days till Christmas. At least here at Lake Hall Park they were away from parents and guardians.

Kate was impatient. Perhaps she had made a mistake coming here again, but she was determined to prise Ned away from the others at some point and provoke some reaction from Guy. Marianne was longing to have a good talk with Guy. She might possibly have a little walk with him away from the main picnic party.

The women soon went in after admiring the fountain which was burbling away fed by a small stream that ran through the woods. Shortly afterwards they went to bed while the men smoked their cigars. It would be quite agreeable if the men were not here at all, Marianne thought before she went to sleep. But she knew that once she was near Guy her thoughts would be quite different.

They were all arranged round the picnic basket in a glade in the woods, except for Marianne and Guy, who had wandered down to the stream that purled away enchantingly at the bottom of the slope. Ned had been leaning against a treetrunk observing Kate Mesure, but had finally sat down beside her. Sarah, who had unpacked the large hamper and distributed plates and glasses, was passing bottles of white wine to René.

'They need to be cooled,' he said. 'I'll take them down to the stream for a moment.'

Kate was nibbling a tiny sandwich under her veil, which was most becoming, though its purpose was to avoid any hungry wasps that might mistake her for a pot of honey. René went off to the stream with the bottles as Sarah unpacked the ratafia biscuits and more sandwiches with tiny slivers of asparagus. Aunt and Uncle Leverton had been invited to the picnic, but had said they thought picnics a waste of time for anyone over forty. They were at present in the garden under the shade of a yew tree with their friends.

René returned with the bottles after about twenty minutes. He had gone downstream for he had caught sight of his friend Guy in earnest conversation with Marianne and had not wanted to interrupt them. Shortly afterwards, they reappeared. Kate looked quizzically at Marianne and noticed she looked rather upset. Guy came and sat down by her side causing Ned to move a little way and concentrate on his sandwiches. He was cross with himself. What was he doing wasting his time down here in Oxfordshire when there was plenty to do up North? René unstopped the champagne and Sarah poured it recklessly into the special glasses they had brought with them.

'Shall you give a toast?' René asked.

'To youth and beauty?' enquired Kate lazily.

Guy took up the suggestion, looking at Kate under his long lashes. 'To English and American girls,' he said and bowed. They all sat on after a second glass had been poured and the ratafia biscuits and the fruit consumed.

'I shall walk by the stream,' announced Ned. 'Will you come, Kate?'

'Oh, I am far too comfortable here,' she replied.

'It's cooler under the trees,' suggested Guy.

'Who is to be butler and footman and clear up?' asked Sarah, feeling she had done her bit. Ned, who had got up to stretch his legs, began to sort out the cutlery and glasses and plates for her.

'We could all walk on a little?' suggested Marianne. 'Take your parasol, Kate, and you will not get burned.'

'Oh, very well, I suppose I must – just to show you Americans are not lazy.'

But no one made a move till Sarah stretched up again and put out her hand to René. 'Come, cousin, let me show you some English flowers. The wood is carpeted in spring with bluebells – too late for them now, but there are wild orchids and many unusual ferns.'

Slowly the others arose, the men helping up the girls and they all passed through a further glade in one happy band with the warm smell of pines and ferns and the sun shining through the green gloom tossing coins of flickering gold on the paths.

Kate and Ned seemed deep in conversation and Marianne and Guy had walked on ahead once more. Unseen by the others he put his hand through hers. The champagne had affected them both a little and she felt happy and attractive. She darted him a quick glance before taking the hand away, then after a moment returned it. The others were nowhere to be seen; they had wandered away from them towards a bridle path that cut through closely planted trees. It was darker under them and her dress caught on the brambles underfoot.

'Don't let's go any further,' pleaded Guy. 'It is hot – and I would rather contemplate you than the trees.'

She stopped and took her hand away again and leaned against a giant oak. He put out his hand and touched her face and she felt a sharp jump in the region of her heart.

'English girls are so lovely,' he murmured, 'and soon I shall be gone. What will you be doing, Marianne, this time next week?'

'I shall have gone home, I expect. It will be dull,' she said in a subdued tone.

'You will have all the young Englishmen asking for your hand – nice, tall men with fortunes,' he said teasingly, but a little wonderingly, for two tears had begun their journey from her eyes down her cheeks. Was the effect of champagne to change a woman's mood in twenty minutes?

'*You* will be gone, and I shall miss you,' she said. 'Oh, Guy – ' If only he would propose to her now.

He looked at her and said, 'Marianne, you are very sweet.' Then he kissed her. She was as warm and soft as a child, but he could see when he drew away that her emotions were unchildlike. Suddenly she threw her arms around him and kissed him full on the lips. His caution reasserted itself. 'No, no, Marianne, you must not kiss me like that,' he said, and planted another, chaster kiss on her forehead. 'You see,' he said as they walked along still holding hands, 'Frenchmen are not cold like the English. Good girls do not encourage us.'

But you kissed me, she wanted to say. *You kissed me*! I don't want you to go away. 'Why must you go away?' was what she said.

'Ah,' he said. 'You are the sort of girl who needs to be married.

I am not ready for that. I am not rich, you see. In France we arrange things differently for nice girls.'

'Have you known only nice girls?' she asked, emboldened, and slipping her hand from his. He thought of Cremorne Gardens and Mrs Hall's.

'You must not ask that,' he replied. 'Men know what they are doing – '

'I am not a nice girl,' she said. 'Not at all. I love you, Guy.'

It seemed that he was unsurprised, though the word love was perhaps not the one he would have chosen to explain his own feelings for her. '*Je t'aime bien*,' he said.

'*Je t'aime*,' said Marianne. 'And I can't bear it, that we must part.' She was half panting, half whispering, her colour high, her eyes moist.

'Let us talk of it tomorrow afternoon,' he said. 'When the others will be out riding.'

She agreed and he turned and went back along the forest ride, she following him and trying to regain her composure. At that moment she would have done anything for him. He turned round once and smiled and then they had reached the path near the stream once more and Ned and Kate were seen leaning against a fence halfway up the slope, deep in conversation. Guy looked towards them and shrugged his shoulders. Sarah and René were down by the stream dipping their feet in the water like children. Guy went down to the stream himself and Marianne followed him. Kate watched from the fence as Ned decided to join the water worshippers too.

It was clear that Guy did not wish to take off his boots and stockings. Marianne, too, felt shy. Ned simply took off his and waded out into the stream looking at no one. Kate laughed and put up her parasol.

'Someone had better go and fetch the pony,' shouted Sarah. 'I left all the baskets tucked under a hedge while you went gallivanting,' replied Ned.

How childish and innocent it all seemed – a small group of friends sporting by the stream. But when Sarah and René dried their feet and Guy and Marianne had stood for a moment looking at the water and then turned to climb back up the slope, Kate went down to the bank of the stream and stood there with her parasol up. Ned turned then and went up to her. Her face was cool and bland under the sunshade and her eyes amused. 'Do you expect me to get my feet wet?' she said. 'I find picnics very tiresome – unless you are in the nursery. Besides, we have to wear so many clothes. For a real Pan party one must swim naked.'

He blushed and looked at her and her eyes challenged his. His glance dropped. She was tempting him, he knew. She was a devil in fashionable clothes, an Eve who always managed to elude him. Did she really think what she said? Was it planned? Or was it meant as an invitation? He did not know and perhaps Kate did not know either.

'You had better put your shoes and stockings on if we are to walk together up the slope,' she said, looking down at his muscular feet with the blue veins standing out and drops of water on his toes. There was something rather obscene about feet, but they excited her. He had gone blithely into the stream, forgetting her for a moment and now she had him back and the others were gone. 'You may take my hand,' she said when he had pulled on his stockings and boots and stood staring at her with the sun on his face. He recollected himself and took her arm as she said, closing the parasol as they walked once more under the trees towards the others, 'We shall ride tomorrow if it is not too hot, don't you think?'

He knew it was a sort of challenge and said only, 'That would be a pleasure.' His voice sounded strange even to himself, coming up through his throat clogged by the passion she aroused in him. And yet he did not like her. She was beautiful and proud, but he did not like her. If he had any sense he would go away that night.

But Ned did not go away that night. He went to bed racked with desire for Kate Mesure and lay trying to think of something else, trying to forget that silky olive skin of hers and that entrancing voice. Marianne lay sleepless too, imagining that if Guy were ever to come to her she would welcome him. Guy slept dreamlessly while René padded up and down making financial calculations in his pocket-book. Sarah dreamed of winning the Derby with some unknown person at her side on a taller horse, and Kate Mesure, having made her toilette, sat filing her nails and thinking how Ned Mortimer disturbed her equanimity. She chastised herself for going perhaps a little too far in the matter of looks and innuendos. But she banished thoughts of him and slept as dreamlessly as Guy.

9

Sarah had gone out riding before breakfast the next morning and was back before Kate was up. She had gone out wanting solitude, for the continual proximity of her guests made her head spin. It was no good her mooning about Ned Mortimer or making a special effort to be nice to him. Absolutely no one knew – and never would if Sarah could help it – that she herself felt attracted to Ned, had done so ever since he had come in April. Wild horses would not have dragged it out of her. And in any case she had never expected he would like her, never mind want her for his bride the way her father imagined. He just accepted that Sarah was part of the landscape. 'What can't be cured must be endured, old girl,' said Sarah aloud to Star as she rubbed her down on return.

Animals had the best of it, she thought wryly. They sported fleetingly in the spring in an unconscious savage way and reaped affection, or what looked like it, when the young were born. She paused a moment in her work and nuzzled Star. Some birds stayed with the same partners and even returned to them; others were promiscuous in their affections. She wondered why. Marianne's Uncle Lechmere would know, but it was not the sort of thing you could ask a man, even a scholar.

She sighed. Ned probably knew about things like that, but his concern at present was not the mating habits of birds. She laughed a little to herself and thought, I am getting as mad as Marianne. 'We'll be able to go out hunting soon,' she remarked to Star. Just then Kate was to be seen at the end of the yard and Sarah waved to her. She would want the mare that afternoon if she and Ned were to go out hacking together. She shouted, 'I'll remind Osborn about your horses,' and with a last pat to Star, went slowly into the back of the house.

It had been some time since Guy had had a woman, and he was feeling lickerish. The word tickled him. 'Feeling lickerish dear,' a woman in London had asked him on one of the evenings when he and René were not paying court to the ladies of society, but worshipping at the altar of a society of a different sort. England had been an eye-opener. Such abandonment in one quarter, such respectability in another. He had supposed that it would be difficult to find a little friend, this being a country cursed by ideas of the proper and the polite and the puritanical. But his first evening in the West End had been enough to disabuse him. It had been startling, almost frightening, the way the *demi-monde* – the poor little prostitutes, dolly mops, fancy whores and drabs – had accosted him near the Haymarket. In parts of London things went on of which he was sure Sarah and her friends knew nothing, or if they did dismissed them as belonging to a remote other world.

The shouting and the screams and the drunkenness were worse than he was used to in Paris where people of all classes sat in cafés, the respectable as well as the workers and the criminals. London was a mess, he thought. You never knew what sort of person you might meet or whether it would be safe to pay for the pleasure of an hour with a dolly-mop in an accommodation address. The girls were as dirty as the city which lay under a pall of soot, even in summer. He had been told that in winter, with the fogs and the rain and the eternal soot it was even dirtier, and could not imagine it.

There seemed tc be two Londons: the opulent, stately, rich London of Park Lane and Mayfair, the houses where the fact of his knowing René had allowed him to join in the season; and the other London, sometimes only a street away. But the two *did* meet. They met in the way the girls dressed, the way the men drank, and the pranks the rich men sometimes played on the poor when in some drunken escapade they would fight in the Café Roche or even at Cremorne Gardens, necessitating the constant vigilance of the police. He had been to other places too with René and a young Englishman with whom René had formed an acquaintance; to Mott's, where he had been puzzled by what sort of girl accompanied men there. There were educated girls, 'actresses', well-dressed girls who even if they had fallen did not seem to have fallen very far, surrounded as they were by young men, some of whom he had even met in 'Society'. There were older men at Motts too who were Members of Parliament, peers, or captains in the army, with women who seemed always afloat on a tide of champagne.

'They call these women "soiled doves",' René told him the first time they walked down the Haymarket. 'Let's turn into a place and eat oysters.'

'Just off there on the right are Turkish baths disguised as brothels – or the other way round,' said René's English friend.

The Haymarket had dazzled Guy that first time late at night after a dinner in Society. He had confused it with the respectable places where his days were spent and then had seen how very differently the girls behaved on the streets. But were they really all so different?

Later he said to René, 'Surely there is something between the stiff and starchy respectable girls and the little whores?'

René had considered this and said he thought the very rich and the very poor much the same. He was more careful than Guy and perhaps he had not even gone with any of the girls there but just pretended to. Guy had succumbed once or twice, but could not help looking with the eyes of a Frenchman at the display of female charms. To him all Englishwomen were lacking in taste, just as their men could not take liquor without becoming violent or pugnacious. These women in their crude purples and reds and hectic green dresses and their 'gold' jewellery matched the society girls who, he thought, wore dresses like florists' shop-windows and over-flounced crinolines, with too many clashing colours, over trimmed their hats, and floated in veils of gauze more appropriate to goddesses than nice, ordinary young women. But, though he could have helped them with their dress and deportment, he could not afford the nice girls, and was a little frightened of the cheeky street women in their poppy reds or violet silks and grass-green skirts.

'You are fastidious, my friend,' said René.

'They are all so gaudy,' replied Guy.

'Miss Mesure dresses well,' observed René. 'I would wager she knows about life. At least it cannot have escaped her that England is not the whole world.'

'Oh, Kate is sublime,' replied Guy.

But it was Marianne who had asked him that morning about the other London and what men did down the Haymarket and whether ordinary girls could go to Cremorne!

'Of course, there *are* Society girls there, I believe,' he said. 'Not on the streets or in the clubs or rooms, but I saw rich girls at Cremorne.'

She looked thoughtful after that. Men could go everywhere, whilst if you were a young woman you had either to be extremely smart and rich, or fallen to have what they called 'fun'. It was not *her* choice to be respectable – her parents had made her so. And they had always impressed upon her that the slightest deviation from strict morality would bring about ruin. Would they say the same to her brothers? It was no fun belonging to the middle classes. Either an aristocrat or a pauper would have been better. Would she, though, like that sort of fun? She was not sure. She was jealous of Guy's knowing about it certainly and his knowledge also excited her. But she did not like London. Even the white stuccoed houses on Park Lane and the luscious drawing-rooms she had frequented that summer were not really to her taste. Why could one not be a friend to a man, get to know him? Some girls, she had heard, treated men as comrades and even smoked – but they were called 'fast' by her circle. She had tried to explain this to Guy, but he had seemed to think it was not possible for women to be free like men and that in England ordinary girls were either virtuous or fallen. He did not use those words, but Marianne knew what he meant.

'She is a little puritan who would like to be a *grande horizontale*,' said René when Guy reported the conversation. He hoped that Guy did not get himself mixed up with a 'twopenny upright' on Panton Street or Windmill Street. He cared for his friend and it was true that London vice was lurid and rather disgusting. He had already met plenty in Paris, sometimes timid little English governesses who had been 'exported' from the shores of Albion, to be locked in *maisons tolérées* and officially supervised. England, with its little private rooms in shops above the Burlington Arcade (his English friend had pointed them out to him with a snigger) and its crime-sodden alleys in other parts of the Great Wen seemed mistress of contrasts. He had bought a lemonade for a child once in a squalid London tavern and then melted away, not before noticing the amused glances of several *filles de joie*. You had to know your way around and he had summed up London to himself as a painted cess-pit, a Babylon. René would much rather keep a girl of his own – he was fastidious. The sort of girl he liked could not be picked up at the Holborn Casino or even at Mott's. Some of the *grandes cocottes* who had taken over the Ladies' Mile of Rotten Row, which had formerly been reserved for Society beauties, might be more to his taste – but he preferred the Bois de Boulogne.

Guy agreed with René that things were better in France. He too was missing Paris. It would not be long now before he would return.

As Guy passed the open door of the Library after luncheon, he saw Marianne reading there, but she did not look up.

Marianne was reading a copy of *The Times* and learning a good deal from its daily perusal about the society in which she lived. But she was alert to the passage of Guy and could not truly concentrate that afternoon.

Ned Mortimer had never visited a prostitute and his only sexual adventure had been with a farm servant when he was twenty. It had made him uneasy, but he had tried to forget it and also to banish what his contemporaries called 'impure thoughts'. It had not worked. He knew now that he wanted Kate Mesure. It gnawed at him that she treated him coldly, let him suffer, and seemed amused. Like Marianne, he had no insight into her motives, but he was damned if now that he was no longer a callow youth he was to be dismissed as a flirt.

The afternoon of Friday the sixteenth of August was sunny and warm and Ned had promised to accompany Kate on her ride. Marianne was to stay indoors and Guy was in the Long Gallery when they left. Sarah and René were to walk over to a cottager whose child had whooping cough. Sarah had offered to go over with broth and, surprisingly, René said he would be glad of the walk. His cousin Sarah soothed him. He felt she was a good woman and she was undemanding. She would make a good wife.

Aunt and Uncle Leverton were both away on their own pursuits. Aunt Charlotte had gone to town on behalf of her sister to choose new material for curtains in the company of the housekeeper, and Uncle John was snoozing in a wicker chair near the croquet lawn. He so rarely had any leisure that he did not know what to do with it. It happened that Marianne was on the way back to her room from the Library, along the corridor where Hunt the Slipper had taken place in April, when Guy came up the stairs from the other end. The sun came lazily through the blinds of the landing window and they both stopped when they saw each other.

Marianne had been waiting for this moment when she and Guy could be alone and no one would interrupt them, no aunt, or chaperon, or friend. Was it too much to ask? He must have thought the same and knew he would find her up here. For a moment she stood gazing at him in the half-light. What had he said before? '*Je t'aime bien.*' Something in her attitude moved him for he went closer to her and shielded his eyes from the sun which entered the corridor in a thick band of light that had escaped one side of the blind. How could she repress her feelings, just pass by him and not think the moment fateful?

'Were you going out?' he asked lightly and took her hand.

She started. 'No, it's too hot. I was reading in the library – have the others gone?'

'There's no one about,' he said and then suddenly he took her in his arms and they stood swaying till she put her hand behind her to steady her trembling and found the back of the door. She opened her eyes then and looked into his. The pupils were dilated, huge. Oh, she did love him! 'You make me melt,' she said artlessly.

He gave her another kiss, this time prising her lips open, and in surprise she acquiesced, and felt his tongue, warm and probing. She did not know that people kissed like that. She had to say it: 'I love you, Guy.' But he seemed to push her away and she stood distressed, her lips still tender and her cheek rough where his beard had tickled her.

'Don't tempt me, Marianne,' he said. 'You are a good girl.' And he half turned away and this time leaned against the door himself. She was not frightened of him. What could be wrong with love? If only he would say he loved her, even if he did not feel quite as she did.

What did it really mean, a 'good girl'? If your feelings were true and good, why should you be condemned to denial? Some half-remembered words from the article she had been reading not long before swam into her head. It was absurd to hear herself saying them, but she had to say something. She was a good girl, but she was also a feeling girl and he made her legs turn to water and her throat dry.

'It's the middle classes who find love shameful, not the rich or the poor,' she enunciated in a strange faraway voice.

Guy laughed, then: 'You talk a lot,' he said. 'But you *are* a good girl – all your sort are good girls.'

Marianne knew that he was saying something important and that all her training, all her life, all her rebellion, all her reading, all her mother's warnings and her father's hopes were concentrated on this moment.

'I don't know whether I am good,' she whispered, and he came up to her again. There was a light in his eyes she had not seen before.

'Don't tempt me, Marianne,' he said again.

He had decided to 'go on' with her, as they said in England, as far as it suited her to let him. *She* was thinking she would like to be a 'fast' girl – oh, she would. But she didn't want Guy as a comrade. She wanted him as a lover. How did it happen? Was it just once and then a lifetime of memory? No! A marriage – she loved him – he would propose to her when he saw how she could love him . . . There seemed to be a current between them in the space of the half-open door, a current like the magnetism people spoke of, a tug away from respectability and denial.

She pushed open the door and looked inside. It was a deserted dressing-room. The house was half empty of guests and no one was sleeping there. She looked in at the room in the half-light. There was a *chaise-longue* and a cupboard and yellow curtains stirring in the warm late summer breeze. She felt she was in a dream in that strange light. She went into the room and he followed her and shut the door. He seemed also to be in the grip of something beyond his control. Could they sit down on the sofa and hold hands and talk? He looked a little scared now and she was disappointed, so took his hand and went to the *chaise-longue* and sat on the edge with him standing in front of her, regarding her with a sort of fascinated purpose. Then he knelt down at her feet and put his head in her lap and she was indescribably moved.

'Will you let me love you then?' he asked in that husky, accented voice she so loved. 'Just this afternoon,' he added. 'It will be something for you to remember.'

She thought his words strange, but was past caring. For answer she bent her head and kissed his curly brown hair. She felt excited, yet tender, not frightened. Should she be frightened? Then he got up and she stood up too, awkwardly, and he led her to the yellow-draped bed. For one long moment he looked at her, swaying.

'They say in London, "are you good-natured, dear?" She did not know it was what light women said, the women he had had in the Haymarket, and laughed. He realised she was going to co-operate and so threw caution finally to the winds. So as the slight summer afternoon breeze stirred the blinds and as a little clock ticked away, he managed to divest both her and himself (first slowly, then more frenziedly) of the garments that impeded their mutual passion. She was aroused, didn't care if he hurt her, gave herself to him willingly with all the impetuosity of her real nature and felt quite at home and even happy when he seemed to forget *her* to plunge wildly and breathlessly into her body. She moved to accommodate him and kissed him as he groaned and moaned and then was quieter as the bedsprings creaked and squeaked and she only had time to think, It is absurd – I love him, but it *is* absurd, when with a cry he muffled in the pillow, he stopped. She felt a strange warm stickiness between her legs and he gasped, as though it had been his intention all the time, 'I love you,' and then was still. He did not feel very heavy as he lay for a time on her. Perhaps he had fallen asleep? But no, he had not, for he raised himself again and smiled at her.

'Now you are grown up,' he said. Then he said something in French she did not catch, and then, 'You are not – shy? what is the word – coy?'

'I love you, darling Guy – why should I be shy?' she answered and he thought, This one is made for love, but she is too eager and spends herself, dashes at everything. She must learn to calm down if she is to be a good lover. Aloud, he said, 'You would make an *amante habile* with a leetle more practice.'

She said nothing, but lay stroking his hair. It had not hurt her. They had always said that it would.

'*Est-ce que tu as joui?*'

'What do you mean?'

'Did you feel good – like I did? – some women do.'

Marianne had felt more in the desire than the execution, but knew that there was something else, for she had experienced what he seemed to be talking about in her dreams more than once. 'With more practice,' she murmured. Then, 'You will teach me?'

'Ah, *chérie*,' he said and rose to draw on his clothes. 'I told you you are a good girl – even now you are a good girl. You will marry and be respectable.'

She was stunned. But how could she say, I want to marry *you*? He had said he loved her. 'Can we do it again?' she asked in a small voice.

'Ah, *mais je suis un peu fatigué* and the others will be coming in. You will not tell anyone?'

She realised in one dreadful minute that he did not love her, that he had desired her but not loved her. It was she who had tempted him. It was true. 'Don't go away,' she said, sitting up and tidying her hair.

'I was quick,' he said. 'It was *imprévu* – I did not expect it. Next time for you it will be better.' Then conversationally he added, 'You must not love me, you know. I cannot marry you – I am not rich.'

She wanted to say that didn't matter. They could work. They could live abroad. If her Papa and Mama objected, they could wait till she was of age. But all that was obviously so far from his thoughts that she said nothing.

'I will tell no one,' he said again.

'It will be a secret,' she said and tried to stop the tears coming.

He took her hand. 'You are very sweet, but you are a naughty girl. Men are not like women. But now, you know, you will always have had a Frenchman as your first lover.'

My first *love*, she thought.

Now that he had 'had' her, as they said in French novels, he would not want her again. But why not? 'You could come up to the attics with me tonight – we could be together again,' she said recklessly.

He looked amazed. 'Why, Marianne, you must not. We must be sensible. You might get *enceinte* – it would not be right.'

She thought, If I were rich he would marry me. Perhaps I have done a dreadful thing. But it did not seem dreadful, even if he did not love her. 'I won't give you away,' she said. 'A secret.'

Then they both stood up and she said, 'I must go and dress and wash and tidy my hair.'

It was unbelievable. Life went on much the same. No one would ever know. But she must show him again how much she loved him. Even at that moment she would like to have lain down once more with him. Why were girls not supposed to do such things?

Only when she returned reluctantly to her bedroom and made a lingering toilette did the full enormity of her conduct strike her.

She must have been under a spell. But let no one say it was his fault. She had wanted him like that. She had.

The evening was sultry and Marianne could easily have excused herself with a headache, but she told herself, Now or never. I must grow up and show nothing. So she went to dinner quietly and once the meal started found it quite easy to dissimulate, except that she was not hungry. Weren't people usually hungry after making love? Guy seemed to be. He treated her just as he always had which was surprising. Yet he did not seem to be acting at all. She ate very slowly, trying not to show that she had left quite a lot on her plate, and by the time the pudding course came round she managed to summon up enough appetite to allay any suspicions. They all seemed quiet this evening. The Levertons' friends talked to the Levertons, and the young people were left to their own conversations. It was a nuisance servants always being there, Marianne thought. They were the kind of people who might guess what she had been doing. You could never keep anything from *them* for very long. Guy's eyes met hers once or twice, but quickly glanced away, although she thought there had been the hint of an acknowledgement in them.

Guy was, in fact, feeling cross with himself and it did not help that Kate was rather short with everybody that evening. Only Sarah and René seemed normal. By the end of the meal Guy felt restless. It was not that he wanted Marianne again, rather that he was thinking how soon his stay in England would be over and being thereby reminded that there were several loose ends once he got back to Paris which he must tie up. Work for example – and money. He must not get entangled in an unsuitable attachment here – and Marianne was very unsuitable. He thought, I have done her a good turn. She is the sort of woman who will have lovers, always searching for some return of passion. He did not feel at all passionate about her; he was annoyed with himself for giving way to an impulse which he should have had enough foresight to avoid. He thought with a sudden stab of longing of little Henriette and wondered what she was doing and with whom. But he noticed that Kate Mesure occasionally came out of her brown study to look at him meditatively, and he was interested enough to wonder why. He was too much the gentleman to ignore Marianne entirely, but he gave her no encouragement when she glanced tenderly at him.

Ned Mortimer, meanwhile, was addressing all his remarks to Sarah, and René was obliged to make conversation with Kate. It was only after dinner that Ned announced he had received a telegram in the village – he was needed at home and must travel on the morrow.

'Oh, I am sorry you must go,' Marianne said impulsively, and Ned seemed to draw together his brows and look at her as though he had never seen her before.

Afterwards in the drawing-room, where the girls had adjourned, he came up to her. She turned round quickly, thinking it was Guy. Ned stood there rather awkwardly and said with a little bow, 'May I write to you, Marianne? It would give me pleasure.'

She was surprised. 'If you tell me about your garden,' she answered at random, thinking he was just being polite.

'Oh, I shall soon be back in London – alas – my father will have various commissions for me to undertake for him. Perhaps we might meet there?'

'Why, yes.' What did he want with her? 'I expect I may go up for a little time,' she said. 'Sarah wants to do some shopping and we can stay with her aunt.'

'Good,' he said briefly and was gone out of the room before either Sarah or Kate could speak to him. He did come down once more, though, after breakfast to pay his respects to Mr and Mrs Leverton and to thank Sarah for his stay before going off in the fly. He shook hands gravely with her and looked round.

'Are you looking for Kate? I believe she is packing,' said Sarah. 'It seems her Uncle Adolphe wants her to join him in Paris after all. I believe they are going on to a spa and will take Kate with them. She had a letter only this morning.'

The house party was effectively broken up with the departure of Ned and Kate and when Guy and René played croquet with the two remaining members of the party that afternoon it transpired that they too were about to leave. 'We must not outstay our welcome,' said René. 'Our own parents are getting restive.' Guy had said nothing to René about his adventure with Marianne, but had intimated that he was bored, and the two had decided to cut short their visit. René had his own reasons. He said nothing of his real plans to his friend whom he thought in rather low spirits. 'A night at Cremorne on the way back, old friend?' was all he suggested. But Guy seemed not to hear.

For the past two years René's mother, whom he had loved as a child, but not liked, and of whom he had been a little afraid (for if she was out of temper the whole household suffered), had been pestering him to marry. His father was ailing; it would put his mind at rest if he settled down; he needed a wife for *she* could not look after him for ever; he must have children; he must take life more seriously . . . Naturally he had known plenty of women and would not allow marriage to interfere with that part of his life, but the trouble was that he had never met any woman with whom he might imagine spending the rest of his life and who would also put up with his mother.

He felt sure it would please his mother if he settled down with an English girl whom she might be less disposed to criticise. He wanted to be master in his own house when his father died, not continually at his mother's beck and call. There was money too which would come to him only upon marriage. He really had waited long enough; he must sort out his affairs.

There had been no young woman whom he had met during this English season for whom he had felt more than a passing interest, but there was one who might do. He had seen what a kind heart his cousin Sarah had; true, she was no beauty, but one did not marry a woman for that. She already had the look of a matron, would be a good mother and devoted to his interests.

Physical attraction did not in wedlock need to be of the overwhelming sort – indeed it would rather be a hindrance to a sensible arrangement – and Sarah seemed to give most of her affection to horses. If not pretty, she was not ugly. It might be an ideal arrangement. He should have thought of it before. His mother could scarcely object to his marrying her half-sister's daughter, he thought, an Englishwoman like herself. Sarah would also bring a dowry – not that that was of overriding importance but it would help, especially if some of his father's capital was settled upon his mother in the form of income for life.

Papa might not die for a year or two, he thought dispassionately. He could wait as long as that. He was in no hurry for the actual wedding once he was accepted. And he thought that he would be. It had been obvious the Gibbses wanted their daughter married, equally obvious that Sarah liked him in a comradely sort of way and that she had had no other proposals. There might be a few maidenly hesitations, and the religious business would have to be sorted out, but he would be doing her a good turn, he thought, at no loss to himself. For Sarah Gibbs would not be a managing sort of woman.

Before he left, René was busy in the library writing a long letter. Marianne saw him there and stole away. Somehow she felt embarrassed. If only Guy would stay. But Guy was engaged in badinage with Kate in the conservatory and their peals of laughter could be heard by anyone who passed by.

Just for a moment Marianne was able to waylay her lover on the stairs and say reproachfully, 'Don't go, Guy – are you cross with me?'

He kissed her hand for answer. 'Cross with you, my dear? Why, no – but I need to replenish the wallet you know, and my Papa will have me home for a time.' There might have been nothing between them.

Sarah came in to Marianne the next day when they had all waved farewell to the Frenchmen. 'Don't be sad, dear – he's not worth it,' she said. Marianne almost told her then, but could not. After a pause, Sarah said, 'René was writing to Papa, you know. About me.'

'About you – but what? Is he displeased?'

Sarah sat down in the little nursing chair that was part of the furnishing of the guest bedroom. 'No, I believe he wants to marry me,' she said quietly, her face averted.

'To *marry* you?' Marianne was stunned. Of the three young men, René had seemed to show least interest in any of the young women.

'Of course, we are almost first cousins,' Sarah went on, 'so they may not approve. But he told me yesterday afternoon when you were writing in your room.' (Marianne started guiltily.) 'He thinks we would be an admirable pair. He has been considering it all summer.'

'But surely – ' Marianne was about to say, 'You don't love him,' but perhaps she had not noticed? She knew she was not very observant, and Sarah might, for all she knew, be delighted.

'I can't give him an answer immediately. I asked for a year,' said Sarah. 'He agreed. There is no point in rushing it. At the moment I can't see myself accepting him. I like him and he will be a friend. I have always *liked* him. But I don't want to live in France. Though he has promised me a hunter – says we could live at Neuilly near the Bois and I could ride. Don't look like that,

Marianne – so shocked. I have had no other offers. At present I think I would rather *not* marry, but I have said neither yes or no.'

'You will have hundreds of other offers,' Marianne said finally. 'You must marry only for love, Sarah.'

'No, Marianne. I believe love can grow if one wills it to – provided one has enough in common. But I am not thinking about it for the present. I just wanted you to know.'

Of all the three of them for it to be Sarah who had had a proposal! Marianne was upset. It was as shattering as Guy going away without another word of love. Of course she would write to him. But she was hurt to the depths of her soul that he could not have seen his way to giving her at least a little hope.

10

Sarah took Marianne back with her to London when her parents returned a few days later. The Scottish weather had been very inclement. Aunt Leverton was glad to have the girls' company for a week or two while her own dressmaker measured and sewed and fitted her niece. There was shopping to do and plenty to see in London even though the season was over and only the less than smart in residence. Uncle Leverton disdained this. He had his business in the City and that was that. In the evenings the girls played a pool of commerce or écarté if Uncle Leverton were not at his club. It was pleasant and dull, but Marianne felt increasingly content to have a little breathing space. Every morning she looked out for a letter from Guy, for she had impressed upon him that she would be at Sarah's aunt's – but nothing came. They switched to playing loo and vingt-et-un and even whist and Sarah made no more mention of her offer from René. She had one interview with her father and stated that she wanted to wait before committing herself to anyone. Sir Theodore had agreed – he had nothing against his wife's nephew, but surely his Sarah could do better than that? What about Ned? Aunt Leverton pointedly refrained from mentioning the subject.

Marianne spent the time when she was alone diarising and weeping quietly and copiously. Sarah knew very well, although she did not know the half of the matter, that Guy Demaine had been frightened of Marianne's passion and feared to get himself involved deeper than was politic.

They said goodbye to Kate, who appeared rather distant and *distraite* as she was borne off in the boat train with Uncle Adolphe to Paris.

'I expect Ned did ask for her and she was angry,' thought Sarah. 'She thinks everyone is after her money,' she said to Marianne later.

Privately Marianne thought Ned too good for Kate. It seemed she herself might be the only one who had not had a proposal. Sarah was about to say that she thought Guy not good enough for Marianne, but stopped, seeing the expression on her friend's face, and knowing she missed him.

'You know how I feel about Guy. But I have no fortune and he knows Papa would not allow me to marry him. If I waited until I am of age – and I could wait – we could marry then,' said Marianne.

He is a dilettante, a charming one, but a dilettante even so, thought Sarah.

But on the third day of their stay in Westminster Square, two letters did come for Marianne. Not from Guy though: one from her parents and one other. She opened them at the breakfast table, laying aside the letter from her mother, which she knew would contain more admonitions to return home now that the season was over and it was her sister Eliza's turn to begin with Madame Duplessis. Letters from home always made Marianne depressed. She had nothing to report to them, only her inability to receive a decent proposal of marriage. There had even been the suggestion of a friend of her Papa's who was looking for a wife: an up-and-coming businessman of thirty-five who had been enquiring about the date of her return. Marianne shuddered and opened the other letter. She did not recognise the upright, clear hand with few flourishes on the envelope and opened it thinking, perhaps Guy has got a friend to address a letter to me. Everything she did or thought seemed somehow connected with Guy, but the letter was franked with an English stamp, not a French one. She opened it. At least it was not from Mama.

Sarah watched her as she read the short communication on a large piece of foolscap. Someone writing from an office.

'It's Ned Mortimer. He said he'd write to me, didn't he? Wants me to accompany him next week to the theatre. "Perhaps Sarah would like to come along too?" ' she read aloud. ' "I have been obliged to invite a business partner of Father's and thought this would make a nice little outing for us. Do say you will come. Yours ever, Ned." '

Sarah looked down at her plate. She supposed it was as good an excuse as any if he wanted to see more of Marianne. It was clearly Marianne of whom he wanted to see more. Not her. She would play gooseberry.

'I don't see why he should invite me – it might be rather boring. Would you come Sarah? I can't be rude to him.' She sighed. What had she to do with other men, however pleasant, when she had inwardly pledged herself to Guy Demaine?

'Yes, you must go,' said Sarah quietly. Marianne looked at her sharply.

'Will you come too? He certainly owes you many theatre visits and treats considering how kind your family have been to him. I wonder why he dashed off so suddenly two weeks ago? He makes no mention of it.'

Sarah had wondered too, but kept her counsel. She was sure that Kate had rejected him and and that was why he had gone away. If this were true, his sudden determined approach to Marianne was hardly evidence of a broken heart. If only he had written to her instead.

'I don't feel like going anywhere or doing anything,' said Marianne. 'I wish I were Kate and were in Paris.'

It was during the next week that Marianne began to despair of Guy Demaine. She had written to him, unable to contain herself any longer, but there was no reply. She swallowed her pride (not, she thought that there was much of it,) and asked Sarah to make enquiries from René. After all, he was her cousin and would know Guy's whereabouts. Sarah was initially reluctant to write to René as it might seem to be giving him an encouragement which she could not truly feel but, being Sarah, her kind heart agreed and she dispatched a little note to him at his address in Paris, adding, 'How is Guy and what are you both doing? I suppose Paris is filling up again now September has come?'

Marianne was beginning to have another reason for despair, one which she successfully concealed, even from herself, during the daytime, but it woke her at night. They awaited René's reply and were still waiting for it when the projected visit to the theatre materialised.

Marianne was upstairs trying to decide what to wear when Ned called, so it was Sarah and her aunt who received him. She was putting the final touches to her toilette – though what did it matter anyway how she looked? – when she was called down by Aunt Leverton.

Two men were sitting rather awkwardly on the sofa in the first-floor drawing-room: Ned and his friend, a solid-looking Manchester businessman, Mr Carmichael. Ned rose as she entered and seemed rather excited. He watched Marianne all the way to the theatre in the hackney-carriage, whilst she attempted to make polite conversation about cotton mills and their problems with Mr Carmichael, who seemed surprised she should know anything about the difficulties in Lancashire. Sarah watched Ned as he watched Marianne and was sure now that there was more to this visit than a polite return of hospitality. But Marianne seemed oblivious.

Only in the first interval of the play, a rather tired revival of a Regency favourite, did Ned, who had arranged that he sit next to Marianne with Sarah on her other side, begin to speak to the former, and he seemed nervous. Even Marianne noticed. It brought her out of the brown study which the play had not dissipated.

He began with a little introduction about his work and his father's enterprises, sketched a short exposé of what he had been doing since he last saw her and finally had just time, before the beginning of the next act, to ask her how she was herself. Did she enjoy London? When was she to return home?

'I am glad we could come without chaperones,' he said. 'How can a man talk to a girl with those beady eyes watching?'

'Yes,' she said, a little surprised. 'But Sarah's aunt is aware of your long relationship with Sir Theo – I believe Sarah thinks of you as almost family.'

This was not what he wanted to talk about, but he had no time until the next interval, when they decided to promenade and eat ices.

Sarah was engaged in a rather jolly talk with Mr Carmichael, who was pleasant and *sans façons*, and Ned came up to Marianne with her ice and a determined expression. Marianne thought she had better put a good word in for her friend, for she knew how much Sarah liked Ned. It would be more appropriate for *him* to marry her, surely – not René, whom Marianne felt still to be an unknown quantity. But whenever she thought of René she thought of Guy and her face settled into lines of sadness.

'You look rather glum,' said Ned. 'Sorry, that's rude.' He ate his ice with his eyes upon her and she grew eventually restless under his scrutiny. What did he want to talk about? 'Would you come out with me to another show – or a concert, soon?' he said.

'I think I shall soon be home – Mama wants me back. I have

nothing really to stay for in London. I wish *I* could go to Paris. Kate has all the luck.'

He looked away when she mentioned Kate's name so, in what she thought a more tactful way, she went on. 'I have not heard from Guy, but Sarah has written to René.' She could not tell Ned about René's proposal as it was still Sarah's secret. Ned did not seem to be aware of any special emphasis Marianne put on the name of Guy.

'I expect Miss Mesure will be enjoying herself in Paris,' was all he said.

'I think Kate will always enjoy herself – at least when she is in the sort of company she admires.'

'What's that?' Ned could not resist.

'She is worldly and has a fortune and so is popular. I'm sorry, I sound rather rude, I don't mean to be. But Kate is better suited to France than London.'

'Yes,' he said, meditatively. Then, 'But *you* are very English, Marianne.' He put down his spoon. 'English, and fresh and unspoilt.' He looked quite solemn, gazing at her with his bluff shaven face and blue eyes. He was a nice man. If only it were Guy, though, who was talking to her, eating ices with her. And she 'fresh and unspoilt'! A little lurch of heart and stomach stopped her from answering for a moment, but then she said, 'Oh, you do not know me very well. I too would like to travel and see the world and not grow into a provincial English lady.'

He laughed. 'There is much perhaps wrong with England, but we English are great travellers – always glad to have home to come back to.'

She wanted to say, But what about the poor? What about the way we live now, with our silly season and our feelings always hidden under mountains of ice-cream and badinage and the glare of chaperons. What about our refusal to let people say what they mean? But she did not voice these things, although he might have understood them. Instead she said, 'I know I am rather lazy – not like Kate – but I would like to travel before it is too late.'

'Too late?'

'When you are married and settled and life becomes just like your parents' life, and you know you will never have time to read all the books and hear all the music . . .'

'I should like to give you time,' he said in a low voice and at first she was not sure she had heard him aright.

'I should like to have the opportunity to talk more with you,' he went on. 'Will you let me? Will you let me show you, one day, other parts of the kingdom, allow me to see more of you? Don't look so astonished, Marianne. You must have known that I have come to feel fond of you.'

This from Ned Mortimer who had followed Kate round most of the summer, who had had, it was true, many conversations with her and had danced nicely with her, but whom she had never entertained as a suitor. 'What are you saying, Ned?' she got out, but then the bell went for the final act and Ned said, 'I will write to you about it. Would you let me do that?'

'Why, yes,' she said, thinking, I must tell him about Guy now before it is too late, but how can I? How can I? Something had maybe gone wrong with Ned's life that was not just a turning away from Kate's coldness. She was not under the illusion that he loved her. Of course not. There was something behind it. She must tell him now, that she was far from fresh and unspoilt. He was an honourable man, in some kind of fix.

'There are things I must say to you,' he whispered.

'Write to me then,' she replied and saw Sarah looking at her curiously. She remembered nothing of the rest of the play and the men left for their club after seeing the girls home.

Ned Mortimer was usually a slow-moving man with a deliberate sense of purpose. Now he was behaving in a very uncharacteristic way, for two days after their theatrical visit Marianne received a letter from him which spelt out unequivocally his new designs. If she had not had that cold worm of fear she would have made no effort either to encourage or discourage him. He might even have been a pleasant way of passing time till she should see Guy once more. Ned had always been undemanding and friendly and she would not have expected of herself that she respond in more than a friendly fashion to his approaches, but perhaps it was a sort of fate that was pursuing her, now that Guy had gone away.

She went over and over that hour spent with Guy and cried so many tears for the unrequitedness of her love and the loss of her dignity, that there were no more to cry. If, for whatever reason, Ned Mortimer wanted to court her, she was too tired to rebuff him. And as a week and then another week passed and it was a month since her afternoon with Guy, she began to see Ned as a hand held out to a drowning woman. But just once more she wrote to Guy. She would give him another week and then make up her mind to allow Ned to go on writing to her or refuse him more intimacies. It was up to Guy now.

Nothing, however, came from Paris and the cold fear that had tormented her began to take shape as a definite possibility.

Ned came round several times, and postponed his return to Westmorland. The first time he acted in a strange way, denigrating himself and speaking of a misspent life, but in terms so vague she could not understand what he was driving at. The second time they went for a walk in St James's Park where he had arranged to meet her. Sarah had agreed to pretend she was accompanying her friend and stayed at a distance while Ned and Marianne walked round and round the lake. It was a beautiful autumn afternoon and the hazy sun and russet tips to the leaves and the peaceful call of the fowl made Marianne feel that perhaps she was mistaken about her fears. Her lightheartedness was not to last, but it sufficed for that afternoon, and she actually enjoyed the walk and listened to Ned in a haze of goodwill. But when he said, 'I have made mistakes in my life but I know I am not making a mistake now,' and they sat on a bench like two lower-class lovers, she felt nothing but a sort of peaceful languour. But then he said, 'May I write to your parents? If I received their permission, would you allow me to propose to you?' and she looked at him in sudden astonishment, mingled with a queer sort of hope. If she could not have Guy, and she obviously could not, would it be wrong to accept Ned? The trouble was, time was short. He echoed her thoughts from far away when he said, 'I do not believe in long engagements, Marianne – I have made up my mind. If you would have me, we could marry immediately, unless your parents wanted a big London wedding and much ceremony. It is unfortunate, but I have to go abroad on business for my father in November till Christmas.'

She cleared her throat. 'My parents have many children and not enough money to give their daughters expensive weddings. And I would not want that. If I married you, Ned, I'd like it to be simply, and soon.' What was she saying? The words were out of her mouth before she knew.

'Oh, Marianne! Then you have given me your answer? You *will* marry me? I promise I will make you happy. You would have to live in the North, you know, but we could travel in the winter if you wanted.' His eyes shone.

'Ned,' she said in a small voice, recollecting herself. 'Do you love me?'

'Oh, Marianne, I think I do – I know we should be suited to life together. I don't expect you to love me, not yet, but I want to make you happy, and then perhaps . . .'

Marianne heard herself saying, 'I thought you loved Kate.'

'No, no!' he cried distressed. 'I promise I never loved Kate.'

But you asked her to marry you, she thought. There must be some flaw in all this. Or why would you be courting me? Yet it seemed a just exchange. She was ruined, fallen – what alternative was there? A visit to an old woman on a back street? Or a plunge into the lake? She looked over at it and shuddered. Or disgrace for her parents and a life abroad as a governess. She could do that. She ought to do that. But she was being offered marriage by a man who seemed to have suffered himself. What else could she do but say yes? If she had had money . . . if she had had freedom . . . if the rules of society were different . . .

She would have Ned's children. She stopped herself thinking along these lines and said, 'I am not perfect, Ned – of course I am not – but I think I am capable of love. You are a good man. *I* am not good, you know. You might find a girl who was more amiable, more practical. I can't believe you are asking me to marry you, but if you are, then give me a few days to make up my mind. There is no harm in your writing to Father. They would be pleased. But I must not marry you for the wrong reasons.'

'What wrong reasons could there be?' he exclaimed.

Oh, Ned, Ned, she thought. Every wrong reason! But it is forced upon me. What else can I do? Tell him, tell him, her conscience murmured, and then he will not want you. Then you could ask him for help and he would give it to you. I am not strong enough, she thought. It was all my fault. I did lead Guy astray. I shall pay for it with the rest of my life. It is not a bad bargain perhaps.

'Then I shall go home and expect to hear from you soon,' he said and took her hand and kissed her wrist under her glove.

'We must go,' she said agitatedly. But if I do say yes to your proposal, she thought, let it be soon, very soon.

'There is no reason for waiting,' he said again as they walked slowly back to the Mall where he hailed a carriage. 'I am not so young and I have enough money, if I work hard, to keep a good house – and money is not the reason for marriage, in my opinion.' He looked happy, relieved from some burden himself, and she looked up at him as he helped her into the cab and saw his honest face. Another woman might treat Ned Mortimer badly, but she would not. What was one deception in years and years of loving – or at least affectionate companionship? He kissed her hand again and she was driven away to where she had arranged to meet Sarah.

She still had not said anything irrevocable and when she returned to Westminster Square she told herself she was behaving immorally. All she had to do was tell Ned and the proposal would fly out of the window. But would it? Ned was not an ignoramus and he did seem to like her. He would think she was a wronged woman and probably help her to adjust to a different sort of life. Not as his wife, of course, but as an independent woman far away, living in some *pensione* with her child. But could she not give him something he wanted? Even if he must never know her true reasons for accepting him, if she did?

'He asked me to marry him!' she said to Sarah when they had had supper and were sitting in the small sitting-room which Sarah's aunt had made over to them.

'I know,' Sarah said. 'I only had to look at your face. What will you do?'

'I shall wait a few more days to see if Guy writes to me. I don't love Ned – you know that – but you have often said that that is not essential for marriage.'

'I said it because I thought you put too much weight upon it,' said Sarah, threading her *petit point* needle.

'If I told him the truth,' said Marianne, looking into the fireplace, 'I think he would rather not marry me. I wish he loved *you*, Sarah. You would be the right wife for him.'

'There is no reason in these matters,' said Sarah. 'In marriage – yes. But we do not choose whom we love.' Marianne knew then that her friend loved Ned and that her feelings were as hopeless as her own for Guy.

All night she wrestled with her conscience. She even prayed. There was no one to whom she might confess herself, no one from whom she might ask advice who would not castigate her and see she was dismissed from decent society. For one afternoon's passion. For loving a man and following her instincts. The world was cruel. Perhaps Uncle Lechmere would understand, but she could not embarrass him with the matter. She was alone and she must choose. It would probably kill Mama if she knew the truth, and the life of her child too would be unpleasant and irretrievably stained before he or she even grew up. The die had been cast. Still, though, she could not bring herself finally to accept him, which she must do if her pregnancy was to seem normal. She would have to be married immediately. Society acknowledged eight-month and even seven-month children. She counted days and weeks on her fingers, sat till dawn wrapped in a rug in her bedroom going over and over her misgivings. The baby might die. Would she still want to be married to Ned Mortimer? Yes, she thought, she would. Love had proved a mirage. What else was there left for a woman?

How could it have happened? Surely it was not *so* easy to conceive? She tried not to think about that hour, an hour spend doing the most wicked thing a young woman could do and yet an action repeated thousands of times in marriage when it was woman's duty, not her desire.

Next morning Sarah came into her room.

'I've got to confess,' she said. Marianne sat up in bed, her eyes red from the night's vigil. 'Ned did ask my advice about you,' she began hesitantly. 'I should not have interfered, but – it was when he left that day at home, when he asked if he might write to you. He asked me if I thought you might accept him. He took me into Papa's gunroom and he said, "Am I wrong to hope that Marianne might like me?" I didn't know what to say. How could I tell him about your feelings for Guy? I tried to be honest. I said, "I think Marianne is capable of love – she has always loved something or someone as long as I have known her." '

It was against Sarah's own interest, thought Marianne. What it was to have such a friend. She had still said nothing to Sarah about her condition. Sarah would not be shocked about *that*. That sort of thing happened more in society than people imagined, and girls were bundled off to the Continent or took the waters conveniently far away. But she would mind that Marianne fully intended to pass off another man's child as her husband's.

'Why should he suddenly take an interest in me, never mind

want to marry me?' puzzled Marianne, glad to change the subject away from her own feelings to someone else's.

'I think it must have been because of Kate,' replied Sarah. 'I suppose he was disillusioned with her and instead of going away to mope he wanted to boost his self-confidence by trying again straight away.'

Marianne laughed. 'You are very direct, Sarah – that is what I like about you.' Apart from this loving business, how much easier it would be to settle down with a woman. Sarah was very restful. They said no more just then. Still she hoped against hope for a letter from Paris – but what would she say to Ned, then? It would be another blow for him. Two girls in a row for him to be disappointed in!

The day passed slowly. She tried to read, as the rain fell steadily on the square, cancelling the promise of the day before when summer had seemed to have been prolonged. She kept hearing music in her head – a tune she could not get rid of – 'Greensleeves' an old folk song she had heard people in the village sing, a man's song sung by women. It was all too appropriate. Had men and women always been so bad for each other? Why had she loved Guy to distraction? Was it just Mother Nature finding the quickest way through the layers of stuffiness and 'civilisation' to carry on her work regardless?

At four o'clock the Leverton footman came in with a silver tray, on it a packet. 'This came for you, Miss,' he said to Sarah. 'Leastways it's addressed to you both.'

Sarah took the packet and opened it. Inside was a neatly folded newspaper.

'It's René's writing on the parcel,' said Sarah. On the top of the double sheet was written a note: 'Turn to page 3' it said.

'It's French!' said Sarah, looking at it doubtfully.

But Marianne took it from her and turned over the page. Then she sat down, suddenly feeling giddy. Sarah took it from her and her eye was drawn to an announcement ringed in black ink. She read it. 'Even *my* French is good enough for that,' she said, after a moment.

'How could he?' Marianne was choking. 'Let me see it again – perhaps there is some mistake?' But no: *'LES FIANÇAILLES SONT ANNONCÉES DE MADEMOISELLE KATE MESURE DE LONDRES ET BOSTON ET M. GUY*

DEMAINE DE NEUILLY. LE MARIAGE SERA CÉLEBRÉ VENDREDI LE 20 SEPTEMBRE A L'AMBASSADE AMÉRICAINE RUE DU FAUBOURG ST HONORE EN STRICTE INTIMITÉ.'

'Oh, my God, oh, my God.' Marianne was saying. Sarah knelt by her, took her cold hands.

'Don't, Marianne – don't.'

But even then in her misery, Marianne's little interior voice was saying, You needn't be hysterical – you knew he couldn't love you. But to *marry* – not to conduct a passionate liaison, which Kate could have carried off and Guy was obviously used to, she saw now. Marriage – to be joined for the rest of their lives. Kate and Guy. Kate, who had spurned Ned. Guy, who had made love to Marianne so ardently. It's her money, she thought. It must be her money. But why should Kate want to marry *him*?

'Did she say nothing before she left – to you?' asked Sarah. It was only a fortnight since they had seen her off on the boat train, waving calmly at them, Marianne with her heart full of Guy Demaine, already with child by him.

'Why did Kate want him?' she cried. 'She doesn't love him – nobody loved him the way I did!'

She realised the enormity of the whole summer's affairs. Nothing had been as it seemed. At what point had she lost herself? What was to be the result of her own folly? I was not thinking at all. What was Kate doing? When did she decide to have him for herself? Will he have told her about me? She could ruin me now. But no, Kate was not that sort of person. She would keep her counsel if Guy were injudicious enough to confess himself. He would not tell her straight away. But one day he would, she thought. One day she will know.

'I told you – Kate always wanted him,' said Sarah, her face scarlet with anger.

'But why? Why?'

'She wants a husband who will be grateful to her, who will amuse her, travel with her, set things up for her. He will have to do as he is told. He is signing away his freedom. I wonder whether he will regret it?'

'She will be sure to have arranged a marriage contract to her advantage.' said Marianne after a silence. The power of money! Yet Kate was their friend. Kate would know nothing of Marianne's conduct with Guy. How could she? She had always teased Marianne about her infatuations. Teased and perhaps warned?

'Yet I *like* Kate,' said Marianne when Sarah said nothing, but looked into the empty grate. 'If I had her money I'd have done the same thing – if I loved him. I just did not know that *she* did.'

'She led Ned on to make him jealous,' said Sarah finally.

'So Ned was hurt and came to me instead,' replied Marianne. 'What a foundation for holy matrimony.'

'You could make it a success,' said Sarah bravely.

So it was Kate who had been at the bottom of it all. Encouraging Ned because she wanted Guy.

'It will make up your mind for you,' said Sarah stiffly.

But Marianne refused to agree that this could be a reason for accepting Ned. She had the more urgent reason. Twice she tried to tell Sarah, but the words failed on her lips even as she tried to articulate them. All she said was, 'If I do accept him it won't be because of Guy marrying Kate. He would never have married me,' she said flatly. 'I was a blind fool.' Why could she not have played games too? Did success always attend the efforts of the devious?

'I wonder what Ned will have to say,' said Sarah, looking up. 'Will you tell him? Or shall I?'

'She has probably sent him the same announcement,' replied Marianne. 'I would not put it past her.'

Marianne was right. Ned called the next day; he did not refer to Kate until the end of his visit when he said, 'I expect you have received the news from Paris? We are to congratulate M. Demaine.' He was pale and seemed disturbed. Marianne wondered again if he had once felt for Kate what she had felt for Guy. If so, it was a just exchange.

Marianne wasted no time. Ned had written to her parents, he told her, and it was now up to her. 'I shall call on you tomorrow before I go north,' he said and took her hand and was gone before she could say anything. There was nothing else for it. Guy would never know about his child now. She and Sarah wrote a joint letter to Kate congratulating her and wishing her happiness. What she must be doing over there in Paris – a whirl of dressmakers perhaps, a marriage contract and then a protracted honeymoon. She must have made arrangements for him to see her as soon as she arrived in France. Uncle Adolphe would have been powerless before that indomitable will. She wanted her fortune now and this was the way to get it.

And Marianne must give Ned his answer and acquaint her parents with her decision. There was no putting off her return home and the speediest wedding that could be arranged. What reason could she give to her Mama? Of course, Ned was to go away on business! He wanted to marry immediately; he must be encouraged, as he had to go away on business for a month or two. Feverishly she made the same calculations over and over again. There was no time to be lost. Kate's treatment of Ned must have led him straight away to her. He thought her a different sort of girl.

Despite her protestations, Kate's marriage to Guy had made Marianne's mind up for her. Her own marriage, she did not doubt, would lead Sarah eventually to accept René. They were all falling like a pack of cards leaning one against the other.

But Marianne could not help thinking that if she had not given herself to Guy, if she had not been pregnant as a result – and why had she not even considered that? – Ned would still have proposed to her and Sarah would still have lost him. Kate would in any case have had Guy. Her beautiful Guy. Would she have accepted Ned without the urgent reason she now had? She tried to think honestly and could not decide whether she would have done in the end. They had had many pleasant talks about gardens and even ideas, though never for one moment had she either flirted with him or desired him. But she did like him, that at least was true.

She dried her tears on Guy's account for the last time and sat down to write a letter to her parents. Ned Mortimer had proposed and if they were willing she wanted to accept him. He was not a poor man; she liked him well; he wanted to marry before he had to go away on business and so did she, etc., etc.

Ned would save her. She would devote the rest of her life to making him happy, she promised herself. She wondered, if the truth were told, whether all young women dashed at marriage like this, even the daughters of dukes and earls, even the poorest women. She was lucky, very lucky. She would make the best of it, Her real life was over, she thought. And Guy would never know, never. The child would have Ned for a father. Yes, her real life was over. And she was only nineteen.

II

Mr and Mrs Amberson were introduced to Ned at Sarah's aunt's in London before Marianne returned home. Their initial awkwardness was soon dissolved by Ned's friendliness. There followed hectic preparations for a marriage on the nineteenth of October in the village church, just across from the manor house which Marianne's father was renting. Her mother did not seem surprised at this sudden match, nor did she seem to have noticed anything untoward. In any case, it was inconceivable to Marianne that her Mama could imagine her daughter might be pregnant.

She thrust down her memories of Guy Demaine with a fearful energy and almost managed to suppress her own knowledge of her guilty secret. She was amazed at herself. But above all she was determined that Ned would never know about it. The marriage was, she realised in lucid moments, only the first step towards the denial of her past feelings.

In fact Mr and Mrs Amberson were delighted that their daughter had made such a suitable match. Neither parent had demurred when Ned had insisted on such a very short engagement. The marriage required a special licence, but Ned seemed head-over-heels in love with their Marianne and a London wedding would have cost a pretty penny.

Ned's father wrote extolling his son to the Ambersons and was to attend the wedding with a few select friends of both families. Mrs Mortimer, Ned's mother, pleaded invalidism and was to remain in Westmorland. Uncle Lechmere had been invited on Marianne's side and he was the only member of her family whom she rather dreaded seeing. She was sure that he would guess the reason for her precipitate marriage. But in the event he cried off, choosing rather to send the happy couple a beautifully bound copy of *The Origin of Species* instead of a cheque. She was to be 'given away' (an expression which she hated) by her father. Ned's friend, Alexander Carmichael, was to be best man.

'Marianne is a dark horse,' said her mother to everybody. 'It's a good match – I can't believe she could have had that much sense.'

Mr Amberson, who was already worrying how he was to find the cash for his sons' school fees for the new autumn term, agreed. 'So long as she's happy,' he murmured. He did not know his daughter well enough to know whether she was happy or not. Although neither said it, the phrase 'Off our hands' was lurking in the back of their minds.

Sarah was to be the one and only bridesmaid, and the bride was to marry in a last season's dress bought hastily in London and was to carry a bouquet of late autumn roses, Ned's expressed preference.

Marianne gave no evidence of pregnancy, was never sick, and did not blush when someone congratulated her on looking so well. She did not feel unwell; when she thought about it she felt both excited and miserable and guilty and frightened, but never let these emotions appear on her face. She was getting used to dissimulation. She was to stay after the honeymoon with Ned's mother in the country while Ned finished his father's business abroad. This she was rather dreading. They were to move in the New Year to a smaller house near the lake. Ned spoke of his countryside in rapturous terms.

There was no time to rehearse the actual ceremony which was to be kept short at the request of the happy couple, and indeed Marianne was astonished at the speed of events once she had accepted Ned and her parents had so readily given their consent. She decided the reason they were pleased was because Ned was gentry in their eyes. She readily gave them the impression that the romance had been going on since the beginning of the season in May.

'The Earl of Durham's daughter was married only two weeks after the announcement of their engagement,' Mrs Amberson told her friends. 'It seems quite the thing nowadays. Of course, Marianne will not have to worry about finances, so there was no point in waiting.' She hoped Ned would not change his mind. Marianne was a funny child, had always been a mystery to her. She was rather quiet now when one would have expected her to be bubbling over with joy.

Sarah arrived the evening before the Day and found her friend calm. Sarah took her duties seriously. 'I think Ned was frightened at going to Europe unmarried in case he met Kate again,' said Marianne coolly. She was sure Kate had been his real love, probably his first love, and he must be feeling bruised. Ned himself had never mentioned Kate again after the time in London when he had proposed. But she was sure that Kate was still his real preference, whatever he said, and it gave her some satisfaction to think this. Each had loved another and so the score could be drawn even. I do wish I could tell her why I am marrying Ned, Marianne thought when she and Sarah sat together in her old bedroom for a last friendly spinster talk. But she still could not. She promised to tell her one day when Sarah was married herself. Poor Sarah who had loved Ned and yet was to be bridesmaid at his wedding. Yet she seemed delighted to have been chosen by Marianne, 'I'm sure to drop something,' she said gloomily. 'I'd better try on my dress. Your mother insisted I should wear pink, though it's just not me.'

They had already inspected the cream satin dress Marianne was to wear. Marianne had been almost ashamed of showing it to her friend, not because she thought she did not deserve it, but because she did not take a great deal of interest in clothes and to wear something once seemed an awful waste.

'You could dye it and use it to dance in,' suggested Sarah.

'Oh, I don't expect there will be many balls in Westmorland,' said Marianne. What did she know about Westmorland or the social life there? Nothing.

Sarah laughed. 'I expect it will be quite an adventure,' she said kindly. How well she would have fitted into country life herself, Marianne found herself thinking. She was sure that if she had refused Ned he would eventually have asked Sarah, to please his father if nothing else. Marianne went to the window and stood looking out at the autumn garden.

'Are you having communion afterwards? I forgot to ask,' said Sarah, tenderly touching the dress which was hanging on a rail.

Marianne had not stayed for the sacrament ever since her adventure with Guy, and intended to miss it until the baby was born – or for ever if Ned did not mind. Her religious faith, which had never been strong, seemed to have evaporated entirely. Her parents were not religious either, except in a conventional fashion. She had

had a long talk with Ned about it. Not, of course, the reasons for her not wishing to take communion, but a proper talk about their beliefs. Ned would not have thought to discuss such things if Marianne had not brought them up herself, but he had confessed then that he tried to follow the Christian ethic, but that he too had lost his faith in the divinity of Christ.

'You are the first girl I have known who is not a believer,' he said.

'Will it make any difference? I will do as you wish, Ned.'

He was touched that she seemed to lay the responsibility for her conduct upon him and, in truth, Marianne was feeling unaccustomedly desirous of doing the right thing. If Ned had said she must turn Muhammadan she would have agreed. If one believed nothing it was quite easy to do as others wished. She thought much the same of this marriage at the bottom of her heart, but determined to try to banish such thoughts. The only thing that troubled her a little would be if her new husband discovered she was not a virgin, but she thought, rightly, that he would not know enough about women to be sure and she could always manufacture suitable reactions.

'No,' she said to Sarah, 'I shan't take communion – Ned did not insist nor did Mama.'

Sarah said nothing, but busied herself laying out Marianne's dresses and travelling cases.

'Don't bother. Taylor will see to it. Sarah, I wish we could just go for a nice walk and perhaps get married in the fields tomorrow. They say some people marry in their own homes – they don't always need a church.'

'Let me brush your hair,' said Sarah.

'All right, it always feels nice when you brush it. Do you think husbands are expected to brush their wives' hair, or shall his mother's maid do it?'

'Why is his Mama not coming to the wedding?'

'Oh, she has not been well and fears the journey. She has written to me – a delightful letter. I am rather dreading meeting her though. But that will not be for a month when we return home.'

'And Ned will be off again?'

'Yes, but only till the New Year. There will be so much to do. I have had to pack one lot of bags for the journey and another for the North. Mama wants my room for Fanny now – it's strange, this room will not be mine any longer.'

'Does it make you sad?' asked Sarah calmly brushing Marianne's long hair and twisting it up into a knot.

'Yes, I suppose, a little, but I haven't lived here properly for so long – what with the season and your place and Mme Duplessis. It's more than a year since I was here having scarlet fever. I was sure I'd die in this room!'

'Oh, Marianne!'

'Promise me you'll always be my friend whatever happens.' Marianne cried, turning round.

Sarah dropped the brush on to the coverlet and put out her hand to Marianne. 'Of course, and you will be mine.'

'Whatever happens,' said Marianne again. Oh, if only she could tell her! But no, not till after the baby was born. Sarah seemed to have noticed nothing. By Christmas though she would be over four months pregnant. All she had to do was to get herself married, get it done. Then . . .

In the morning she was all animation. 'I must dress and sort everything out or I shall be late. Yes, Taylor, come in, I'm getting dressed.' The maid entered and Marianne, helped by her and by Sarah, was fixed into the stiff, creamy crinoline and the veil.

'You're a picture, Miss,' said Taylor. 'They'll all be looking out for you in the village.'

'Are my flowers in water? And the ribbons? Is the trunk in the hall? Who is helping Mama to supervise the breakfast?'

'It's all being taken care of, Miss Marianne. You're just to concentrate on getting yourself married.' The maid giggled and Marianne felt as though she herself had already drunk two or three glasses of champagne.

Sarah went off to dress herself and a sudden hubbub in the garden was found to be the voices of all Marianne's sisters and brothers – Eliza and Fanny and Tom and the little ones – for once all neat and tidy, and anxious to walk to the church over the fields.

'I wish I could walk there with them' said Marianne. 'Just go to the church and find Ned waiting there and then have a quick blessing.'

But at eleven the carriage came for her with the horses dressed in white ribbons and rosettes and she stepped in on the arm of her father. People were all very kind. Everyone seemed to like weddings. If only they knew! A few villagers were waiting outside the little church and tossed their hats in the air when they saw her. Marianne felt she was in a dream. Only Sarah helping her with her train and veil seemed real.

Once the service started she found herself seeing and hearing it all as though she were taking no part but was an observer with the rest. She had firmly refused to recall her problems and began, in spite of herself, to find the process of a wedding mildly interesting. But she must not delve too deeply into what she was doing! Guilt would return – she knew herself well. For the present Ned seemed overjoyed to have her as a his bride and she was humbly grateful. She knew she was doing a dreadful thing, but why was it dreadful if society forced girls to such expedients? Don't blame society, she admonished herself sternly, even as the vicar was enjoining her to be fruitful and multiply – after all, that was something she would certainly carry out. Perhaps Christians would rather a woman married a good man even if he were not the father of her coming child? Nothing was said about the past in these public words. Nothing was said against the sort of love she had already felt for Guy Demaine. It was the future that mattered. Anything else was between her and the god she did not believe in. A god who was as far away as Guy.

'Wilt thou have this man to thy wedded husband . . . obey him and serve him, love, honour and keep him in sickness and in health, and forsaking all others keep thee only unto him, so long as ye both shall live?' There was something noble about it that quite surpassed her temporary troubles. She offered up a prayer as she said her 'I will' in a distant voice. Ned spoke quietly and sincerely and she wondered what he was thinking about.

Well, she had given up Guy. What was Guy, fickle and uncaring, when weighed in the balance with a baby? Marianne felt everyone looking at her, part of a happy couple, for once in her life the centre of attention. Why had Ned wanted her? She thought about it as the service continued. Would he like her when he knew her better? She felt the congregation accepting them, and the ceremony supporting both of them, conferring upon them that legitimacy that raised them above the 'brute beasts'. Now it felt more real and tears began to prick her eyes. But it was no time for tears.

She would never tell him. She must make sure he enjoyed his first night with her.

Mrs Amberson, who lived her life in a perpetual state of crossness because of the demands on her, too many children and a husband who had not turned out to be what she had hoped for, was thinking what a sly puss her eldest daughter was. She was truly surprised that such a nice and eminently acceptable man had chosen to marry her firstborn, never having credited Marianne with any capacity for reasonable conduct. She might have wished she knew Marianne better and then could weep with the best of them, which was all that was expected of mothers at weddings, but there never seemed to have been any time for that. Just at first when the little milky bundle had lain in her arms in that delicious year before the others had come so quickly upon Marianne's heels, she had savoured her, loved her. But then the child had been an awkward toddler with too much independence of mind concealed under her shrinking sensibilities. And there had been also the necessity of making of her husband and her new family a respectable and worthy social unit. She thought she had succeeded, until the crash in '65. Since then they had had to reduce their claims upon the world and draw in their belts. Papa seemed now to be sure that his new business was a success, but she no longer had the bright confidence in him that had accompanied the earlier years of her marriage. She sighed and tried not to worry. At least Marianne looked happy. Thank goodness she was to be looked after by someone else. Even if she was marrying in such a hurry.

At last, after Marianne had been handed over to Ned by her father in a brisk sort of way and they had pledged their troth and she had felt Ned's ring sliding down her finger, which gave her a queer shiver, and Sarah had given her back her flowers and the vicar had prayed that they should live for ever in perfect love and peace, it was done; they were man and wife and Marianne had her veil lifted by Sarah and looked upon her husband, Ned Mortimer. She thought, I shall have to spend my life with him trying to understand him. Please give me strength not to falter.

I suppose I shall have to have many children, like Mama, she thought. She had not thought of that before, being too worried about the baby she was to have in only seven months. Ned squeezed her hand and smiled at her and looked suitably awed when she smiled back at him with the veil lifted and flowers in her hair. Sarah too was moved and thought her friend looked blooming. Ned would make a wonderful husband, Sarah was convinced. Her thoughts went to René who wished to marry her and she dwelt a little wistfully on that until she was recalled by the wedding march and followed Ned and his wife into the vestry, on the arm of Alexander Carmichael, who was looking rather abstracted.

The wedding breakfast had been overseen by the old housekeeper who had come back for Miss Marianne's wedding and she had done them proud. There were flowers from a friend's glasshouse as well as the late roses, and waxy foreign lilies, and the tables were set with hired gold-plated cutlery and white china. Not many presents, except from the immediate family, as they were not well known in the district, but Sir Theodore had sent a set of china and the bridegroom's parents' presents were waiting in the north. 'Clocks,' said Ned vaguely and some old wine glasses and some silver too. The children had been busy making blotters and firescreens which Marianne was surprised about. Eliza, her next sister, still cross that she was not a bridesmaid, was given a tiny pearl bracelet by Ned's father, which somewhat appeased her and Sarah, as chief and only bridesmaid, was handed a pearl brooch by her groom. 'I shall give you *my* present later,' Ned murmured when Marianne handed him her own offering, a book of Tennyson's verse, inscribed 'To my husband on our wedding day'.

Toasts were drunk; Sarah's parents were circulating easily and pleasantly, though they found more to say to Edward Mortimer than to Mr and Mrs Amberson. After the luncheon and the cutting of the wedding cake and the champagne, the atmosphere was one, Marianne thought, of relief. Certainly *she* felt relieved, glad at last to go upstairs and have Sarah help her into her russet travelling dress. Farewells were taken, the last flower petals thrown, and the carriage finally swept them off in the direction of London where they would set off for Paris the next day. The honeymoon night was to be spent in the Grosvenor Hotel. Once there Marianne felt she would be truly married and could forget arrangements and dresses and presents and plans and, at last, let Ned take her in his arms and do with her what he willed. Would it be agony, would he somehow know she was not a virgin? Would she find him attractive? Would she desire him? She tried not to think about it.

The room they were shown into at the Grosvenor Hotel was cavernous and dark. It was also cold, for the fire was unlit. The sunny October country morning of her marriage day had changed into a foggy, chilly winter evening. Fortunately hot water was brought when Ned rang the bell to summon a chambermaid. Marianne went into the dressing-room, where there was a ewer and a hip-bath and large white towels, to wash and dress for dinner. But she was not hungry. If only they could get it over. Ned was suddenly behind her. she felt his hand on her shoulder as, standing in her camisole, she washed her face in the blessed hot water.

He handed her a towel saying, 'I suppose we must dress for dinner. Are you tired?'

'I wish we could eat up here – I am rather,' she replied, turning to him.

For answer he took the towel back and dried her face and neck awkwardly and said, 'I expect it could be arranged.'

Marianne looked at him and debated whether to put her arms round his neck. After all, he *was* her husband. But she must not be forward. She had never seen him in his shirt-sleeves before, never mind undressed.

'You don't mind?' he asked.

'What?'

'I supposed that women did not want to wash in public, even before their husbands.'

'It does not seem to me very important,' replied Marianne, laughing. 'Are *you* tired, Ned? It would be surprising if you were not.'

'London is not a good place to feel at home,' he said for answer and went across the dressing-room to fiddle with the window cord. She had never seen him at a loss before and instinctively took charge.

'I should like a bowl of soup and some fruit and cheese, if that could be arranged. Then I should like to sleep,' she said. 'I want to enjoy our journey tomorrow.'

Ned must have thought that she did not know the facts of life for he blushed, then said, 'I've never been married before! I suppose they would bring us up some food?'

'Try ringing the bell again,' she replied. She put a wool under-spencer over her camisole and sat on the edge of the hip-bath. 'I should like to bathe – it seems a pity to waste the hot water. We should be travelling with servants, I suppose.'

'Thank goodness we are not,' said her husband. 'I shall do as you suggest.'

No one seemed to think it odd that they were not to go down to dinner, and trays of food and drink arrived as if by magic. Marianne bathed quickly while a fire was lit for them and then they sat down companionably to eat, and to drink a glass of wine. It was scarcely a scene of passion and Marianne was even beginning to think they would never get into the bed together, but the wine started to do the trick and Ned began to caress her. So married couples could act just like *demi-mondaines*, she thought confusedly. He looked a little frightened though, as if she would blame him for taking liberties. Whatever was she to do?

'I am tired – let us go to bed,' she whispered. 'Get the servants to fetch the food away and let us have an early night.'

He looked at her doubtfully, as though he could not believe that she was inviting him to share her bed. It is more embarrassment than lack of will, Marianne thought. I must put an end to this. But when the chambermaid and the valet had gone and their clothes had been taken away for brushing and they had let the fire burn low and Marianne had made him ask for more hot water – he closed the door as he bathed – she lay waiting for him to come in to her, thinking of the prayer-book and the service and almost dead with fatigue. He was at her bedside.

'Shall you sleep now?' he asked and took her hand to his lips.

'If you wish,' she replied meekly.

He blew out the candle and got into the bed beside her. For a moment he lay stiffly at her side and then she whispered, 'I am cold – can you warm my feet?'

Soon they were embracing in a tight knot and Ned had decided to forget his shyness and to prove his ardour. Some men may like a resisting wife, she thought – it might make them more excited – but I don't think Ned would, he is too sensitive and would think he was hurting me and would desist. On the other hand if I am too saucy he will think I am experienced, which is not really true. However did I manage to make love to Guy so easily? She decided to act tenderly, but with enough warmth to encourage him. Ned groaned and was soon returning her kisses. At last. She felt his solid, smooth body lying on hers and gasped a little, more from difficulty in breathing than from passion. She put her arms round his neck and now that their eyes were accustomed to the darkness, she saw his looking at her.

'It would be more comfortable on my side,' she whispered and snuggled close to him again. This time he decided to be more bold. Was he going to ask her whether her Mama had told her of the mating of humans? She hoped not. 'I will do what you want,' she said in a small voice and this seemed to inflame him. At last her shift was up round her waist and Ned was ploughing a furrow inside her after a small initial hesitation when even Marianne forgot for a moment that she was not a virgin. He began to move more and more rhythmically, breathing heavily and murmuring, 'Forgive me, forgive me.' What on earth should she have to forgive him for? she wondered. She must respond to him, for was not this the night when their baby was to be conceived? She began to moan quietly with what she hoped he would take for pleasure, but he stopped and said, 'Am I hurting you? I love you Marianne.'

'Of course you are not hurting me,' she said. 'I am quite strong – it is natural, is it not?' And she began to mimic his own movements. This had the desired effect and this time Ned abandoned himself. She slid her legs round his waist to make it easier for him and thought of nothing but his consummation, whispering small endearments. He got rather excited and took more liberties but still, she felt, he held something back. She would not tell him she loved him but murmured, 'Please take me, Ned, now that you have me,' and then he abandoned himself completely and was soon gasping on her neck. She even felt tears on her shoulder and was amazed. She stroked his hair – it seemed the natural thing to do – and found there were tears in her own eyes, but for what she did not really know. Then, 'Thank you,' he said. Somehow she must show him that women had desires too, for he seemed to be under the impression she was doing him a favour.

After that first time he made love to her in the morning again, repeating, 'My wife, my wife,' only to lie back eventually on the pillow and ask her, 'Did I hurt you? I am sorry.'

'Of course not,' she replied and for answer kissed him again.

By the time they had dressed and were on their way to France, Marianne was satisfied that there was every reason for her to have become with child from her husband. Later, in Paris, she was even to wonder whether the increased activity might have dislodged the child already there. At that she felt a little pang. If that happened she would lose the last of Guy. Yet she would still be glad to have married Ned.

She did not lose it, and two weeks later she judged that she might very well now have official evidence of their intimate life together and she told him she might have become *enceinte* that first night, for there was no expected arrival of her monthly visitor. She hated herself a little as she said it, but it was, of course, true and Ned was pleased and proud and even fearful of taking her again in case the visitor was brought on. But Marianne reassured him all would be well – she was sure that she would be a mother next year. He marvelled at her matter-of-factness and wondered whether all women were like this. He had expected coyness and fear and trembling and even a frightened refusal to do his will. But Marianne seemed to enjoy her relations with him and he felt humble and happy. Above all, he felt grateful.

'Of course, one cannot be *sure*,' she said the next week when they were sitting at a café near the Tuileries Gardens. 'But I hope so.'

They looked like lovers now, not a married couple, she thought. And she *was* fond of him. He might even, to judge by appearances really be in love with her. They got on well together, rather enjoyed the same things – walking and talking and looking at Paris, its buildings and gardens and statues. Ned was always worrying that she might be bored, but she assured him she was not.

'I thought most young women liked shopping and buying things. I am a dull dog, I know – I enjoy myself and forget to ask you if you are happy.'

'But I *am* happy,' she replied in amazement. 'I don't want to look for a fashionable life. I should be quite happy to be a peasant – except I should not make a good worker when there is so much to think about.'

He did not make her heart race or fill her mind with dreams of love or make her melancholy. He was an honest intelligent man, if not a dazzling dandy or an ambitious man of genius. But she would not have wanted to be *married* to that sort of man. Ned gave her space to breathe, to be herself, and for the first time she realised that she would never have married so sensibly if she had not had to. As she got to know Ned better she felt more and more sorry for deceiving him.

One night when they had dined and were looking out over one of the new *grands boulevards* with blurred lights waving everywhere from the gas jets, he came up behind her and said, 'I told you you would have my present later. Well, here it is.' He put a little string of pearls round her neck and she turned to look at herself in the glass. 'They are like you, Marianne – pure and singleminded and delicate with a sheen of life.' He kissed her neck and she felt humbled. If only he knew how impure and double-crossing and devious she had been. All she could say was, 'No, I do not deserve them but I love them.' She wondered if he had ever given Kate Mesure a present. They had not come across Kate or Guy or René in Paris. Marianne had at first been terrified that Guy might cross her path. Guy who did not know he was to become a father! It was absurd. She must not be frightened.

'I don't want you to go away – I wish you could come back to England now for good,' she said to Ned at the end of their holiday.

He was to go on to Germany and Vienna after seeing Marianne off on the boat train to Boulogne. It was a measure of the distance she had travelled since her wedding that she genuinely wished he were not going away and knew she would miss him. They were growing together like old friends and companions. It was to be a successful partnership. That she was sure about. And Ned seemed to feel it too. She knew he was proud of her in her white satin evening dress and her pearls, and that he, too, was happy. It was as though some burden which he had been carrying had fallen for him. He was a man ideally suited to marriage.

12

Kate and Guy Demaine at first rented an apartment in Paris near the Luxembourg Gardens. But Kate was restless and planned to return to Italy for Christmas and to stay there – buy a villa perhaps. Guy was happy to agree with her. Their honeymoon had been spent in Florence and Venice and he had thoroughly enjoyed himself. He was in thrall to the glamour of Kate, but not fully aware why she had married him. He amused her; they both enjoyed travel and architecture and buying pretty things and eating and drinking well. Guy had at first thought they might both be under Kate's Uncle Adolphe's thumb, but it appeared that American ladies were given a long rein of freedom by their relations.

Kate, by the intelligent stroke of marrying Guy, had secured herself part of her fortune (the rest was to come to her when she was twenty-five). She was now free to spend money and do as she willed.

In fact Kate had married Guy because he hadn't a bean, was neither American nor English, was 'artistic' and a good mixer and would be forever in her debt. No nonsense about her money being made over to him as would have happened in England. The money was Kate's and if she chose to give him an allowance, that was sufficient. In France, with *séparation de biens*, the marriage contract secured her freedom, and Guy's, so long as he was pleased to be her husband. Of course she knew Guy found her attractive too, quite apart from the money. But money made her even more attractive and his status too was enhanced by her fortune. Even his face and figure looked different from what they had been in the summer, or what they would have been if he had married a penniless Marianne.

It seemed to Guy that two out of the three girls he had met in England had been mad about him and this flattered his ego. French girls had never, apart from little Henriette, been so keen on him. It must have been that he was foreign and therefore new and unusual. He knew he was good-looking, had always been sure of his masculine attractions. But Kate did not seem to be exactly besotted with him. She could be quite cool when she wanted. She found him an asset in public, he guessed, and helpful when she was in what was to her a foreign country. He determined to assert himself in the matter of practicalities. It might be her money, but he knew the best ways of spending it, knew where to find the best carriages, the best wine, the best food. This entranced his wife. She could not have borne a husband without *savoir-faire*, and *savoir-faire* was one attribute that even his worst enemy could not have denied him.

Kate had quite enough to do at first, principally shopping and visiting dressmakers. The dreams she had had of buying gloves and parasols in the rue de Rivoli were now realities. Not only gloves and parasols: dresses, hats, ribbons, jewels. And for this she needed an admiring audience, which Guy was. 'English and American men are not interested in clothes,' she said.

It had been a struggle, getting her uncle to advance the money she was to have after her marriage, before she wed Guy at the American Embassy. But as usual she had got her way. Uncle Adolphe had been sweet; he had at first tried to get her to reconsider her marriage to this handsome penurious Frenchman, but she had looked at him coolly and said, 'But I must marry him,' and he had said nothing after that and gave her a cheque for a thousand dollars. She had signed scores and scores of papers and Guy had had to sign many of them too. Then there was the search for a small apartment. He had thought it would take weeks, but the power of money was amazing. Some American friends came up with an address and before he knew where he was Guy was living in his wife's little apartment. 'Only a *pied-à-terre*,' she said. 'Or don't you call it that in France? Like "cul-de-sac" – a word we Anglo-Saxons appropriated?'

Guy's family thought he had fallen on his feet, and so he had. But he was still, beneath the excitements and the travel and the wealth, a little uneasy. It was the *idea* of Kate that attracted him. She did not seem to be a sensual person. 'My weakness is for strong men,' his new wife had said to him on their honeymoon, and he had thought it was a compliment until she had added, 'It's a nuisance, for I like to be the strong one myself.' She had looked really upset as she said it. They had been breakfasting in bed and perhaps she had been half asleep and the words had slipped out of that half-dreamland that accompanies the first half hour of a lazy holiday awakening. But each day was to be a holiday now! Kate told him that when they returned to France she intended to look for a better apartment near the Bois and move into social life. 'This year is just a little rest,' she said. 'We shall have to work quite hard afterwards if we are to cut a dash, you know.' Guy rather enjoyed the idea of accompanying his rich wife in the life she longed for. But American women had such free manners.

'You will not like having to subdue yourself to French society,' he said in an unaccustomedly caustic tone, and she looked at him underneath her eyelashes. 'You want to be free, dear Kate, but I can tell you that this Second Empire of ours is boring and bourgeois and Catholic – I prefer Italy.'

It had been only a few days after meeting her again on his hurried return to Paris with René from Sarah's house that Kate had made it clear she wanted him. Kate, however, was no Marianne. He had said nothing about his adventure with her friend to his wife. He had enough sense to keep it to himself for the time being. Might be useful later, though, if Kate needed to be made to feel jealous. Somehow he had found himself embarked upon compliments that were not rebuffed; then upon a rapid courtship which he had at first only taken half seriously. Then Kate had led him to propose to her and had immediately accepted him. Why had it not seemed odd at the time? *Was* it odd? Never mind. Here he was now with a wife as rich as she was attractive, the woman who had spurned him throughout the summer and put him on his mettle. And he looked forward to the arranging of their joint social life with a good deal of pleasure. That was what he excelled at. They would be the greatest givers of parties people had seen for a long time, he and his wife. His *wife* who was a mystery to him and perhaps a reward (though for what he did not know). '*You* will let me be free, my dear,' she had said to him one day not long before their return to Italy.

'But you *are* free, *chérie*. Money makes people free, even if you insisted on staying in Paris and mixing with the bores.'

'Free,' Kate murmured. 'Yes, you have freed me.'

'And I am enslaved,' he replied gallantly and laughed.

They got on very well, he and Kate. Sometimes though, she seemed to him to be forty not twenty, and her temper seared him occasionally. It was as cold as ice, unlike any woman he had ever known.

They continued to shop and to honeymoon and slid into a pleasant and amused companionship. Then they packed their trunks and bags again and went off to Rome via Florence, leaving the Paris apartment under the eye of a housekeeper and her little daughter until the following autumn. But other events then conspired to keep them in Italy.

'He can't have been a tuft hunter,' said Sarah's father to his wife of Guy about the same time.

'He hasn't a bean, Susan says,' replied his wife. 'Do you think my nephew is one?'

'Perhaps. Or he's a snob like your sister,' said Sir Theo with a chuckle.

Lady Penelope was used to his strictures upon her half-sister Susan. 'Anyway, Sarah has not decided yet. There is plenty of time. Wouldn't have wanted her to marry Demaine though.'

They both fell silent, thinking of whom they would have wanted Sarah to marry, now tied to Marianne Amberson. 'All for the best, I suppose,' said Lady Penelope vaguely.

'Well, with Miss Mesure's money and the young man's charm, I expect they'll enjoy the fashionable life. Wouldn't have done for Sarah.'

'No, Miss Mesure is the managing kind. She could have done better than Guy though.'

Sir Theo, who had also fallen a little under Kate's spell in the the summer, had concluded that she was probably less experienced than she looked.

'She dressed so well, and seemed so sophisticated,' said his wife.

'She looked old for her age, I thought,' he replied. 'Nice trim figure though.'

Lady Penelope tried to pretend to be shocked, but did not succeed.

'Sarah's fatter nowadays, don't you think?' he asked. They both thought of their dear daughter who had seemed rather miserable this autumn after her friend Marianne's marriage. Sarah would never be a worldly success, thought her mother. Girls often turned out like their mothers. She wondered what Kate's had been like.

'We ought to have had Sarah painted in the summer,' said Sir Theo. 'Those Cabinet portraits made her look fuzzy.'

Sarah was at that very moment out hunting with Star. She was having a lonely time and riding seemed the only thing to cheer her up. She could not make up her mind whether to accept her cousin René's proposal. 'I am bored because I am a boring person,' she said to herself. Bored by society; bored with having come out, yet still expected to enjoy the endless unremarkable activities of an unmarried woman. Driving with her Mama in the afternoon to leave cards at neighbours, listening to tittering girls, having her hair brushed every day, endlessly fulfilling the mechanics of ordinary living, listening to flirtatious men (not many of them now), listening to dull men, of which the countryside seemed to be filled. Bored with praise for her riding; bored with the endless speculations of neighbours and friends as to the marriage prospects of their daughters. Was she even bored by her beloved countryside when she could walk there alone and observe its hidden life? No, not really. But she wished she could take some part in it all – wished she could make gardens, plough, have a place in it. Did she find even her Papa's conversation boring when she accompanied him on his constitutional? A little. But she was even more bored at her aunt's in London. And more than bored by London itself. Sarah hated the greasy four-wheelers and the din of carriages and hansoms on the roads and the sight of the low, black houses and the gin shops. She loathed the livid smokiness and the noise and the dirt and the suppressed, or sometimes not so suppressed, violence of the London streets. London was overwhelming – a dreadful city, hideous, vicious and cruel. The sight of the poor filled her with horror. But what could she do about it? Even her one pleasure, her riding, belonged to the days of childhood. Had she never grown up? She sometimes ate too much or caught herself wishing for a glass of champagne to cheer her up. She had hearty appetites and there was nothing for her to satiate them with. She must force herself out of this rut.

Apart from her riding, the only good things about her life that winter seemed to be her quiet visits to the little church in the village across the park where she taught in the Sunday School and helped with the flowers. She worshipped there every week, but wished for more guidance than the curate seemed capable of giving her. She felt her laziness would disappear if she had real work to do. The trouble was within herself. Ought she to marry? And, more to the point, ought she not to decide soon about René? She sensed that she would not be bored by him. But René was nominally a Catholic, and she would not convert. That was not the only stumbling block and it could perhaps be sidestepped; she felt that René did not take religion very seriously. No, the trouble was that she sometimes woke from dreams of Ned Mortimer. There was something she wanted that she would never have. Perhaps it was allowable to hope for it from René if he were a husband?

One November day she received two letters. One was from Marianne who was now in Westmorland. She sounded meek and un-Marianne-like. But her great news was that she was expecting a baby already. Dear Marianne – and dear Ned. What good news!

The other letter was from René himself. He had just seen Guy and Kate at a little soirée they had given in their apartment before going off again to Italy. They both appeared very well and were enjoying themselves. But René informed her too that his father was ill and because of this he begged Sarah to reconsider his proposal. It would be so good if his dear father – were he not to recover – might know that his son's future was assured if Sarah accepted him. It would be enough for him to know – they need not marry immediately.

This letter made Sarah feel uneasy and she returned to Marianne's once more. It was odd that Marianne was now enjoying the sort of life which so bored *her* – perhaps being married changed one's perceptions? Sarah was not usually so introspective and began to wonder if she ought perhaps to consult Marianne the Married Woman about René. Now when did she say the birth of the baby was to be? Ah, the end of July. Gracious, she must have conceived on her honeymoon. How exciting! Sarah went into a reverie before she returned to her friend's letter once more. There was a long paragraph about France, which Marianne had loved. She praised the French way of life to the skies – everyone always ready to talk about things in general, just the sort of conversation she loved, with people erecting theories about everything. Ned, too, had been more talkative than she had been prepared for and, although English to the core, had enjoyed the change from dirty London to elegant Paris with its gleaming new boulevards. Just as her life was beginning to open up, Marianne hinted, the return to England had closed her in again. But she would feel more cheerful when Ned was back. The countryside around his parents' house was magnificent and made up for missing Paris. How could one like two such different places? Yet she did. Mrs Mortimer was arthritic, in a good deal of pain, but had been very kind to her. Ned took after her – but Sarah would know all that. Her parents must often have talked about 'Tubby' and his wife? She, Marianne, was going to make the best of this new rural way of existence. No one in the south of England had any idea of the roughness and grandeur of the North. She loved it – though had not yet been to what they called the more scenic parts. It had been enough to look out of the large sash windows of the Mortimer home at the little house that was being prepared for them at Langthwaite with the lake almost at the bottom of the garden, the distant peaks of the Langdales in one direction, Coniston Old Man on a good day on the other. And the lake was beautiful – no trippers in the winter and everything very cold and pure. She was sure that the air was doing her good and all the wonderful home cooking the staff went in for. And their language! It often sounded like old Norse. And the way they never flattered but took up 'sharpish' as they would say if you said anything silly. The servants were more like friends here. The family seemed to be loved in the village and in the little town, and in short nowhere could be more conducive to harmony than this countryside even when you were so cold that three stone hot-water bottles could not warm your feet.

'London seems like a dream – even Paris seems nearer than that. The people here are not humbugs and are slower to make a fuss about things. I don't know whether it's the country life or the fact that it's the North Country, but I like it. There seems to be quite a lot of hunting, though of course I don't partake – wouldn't even if I were not in a "delicate condition". They praise my complexion up to the skies for they tend to be rather red-faced and shiny with all that wind. I can see that Mrs Mortimer's housekeeper would rule me with a rod of iron if she could. They think I am a child, but they like children. It's true there are few people of one's own sort or age to talk to, but I don't seem to mind that at present. At Christmas there are said to be all kinds of goings on – country dances in the big houses, so I may see some younger folk. I thought I should just have to 'grin and bear it', Sarah, to be honest, for I had no idea of Ned's family or of his place, but now I find I am very lucky. I thought his mother might snub me since we married in such haste, but she is glad, she says, and though she does not yet know my momentous news, I know she will be thrilled. Two things though have made me laugh here. One is the fear of strong drink which the Methodists have brought with them – even a tot of brandy and water is used "only in an emergency" and by some people never, even some of the better-off villagers and farmers. The other thing is the atrocious fashions they all wear – about fifty years out of date. I need not have worried about my wardrobe, and that pleases me, for you know I can't abide fuss over clothes. I wore my little mauve afternoon dress the day the friends and neighbours came over for afternoon tea and you should have seen their faces. I might have been wearing ferocious violet or mustard yellow. They had never seen such a colour, they said. Where did I get it? And the cut was "very fine". Two or three of the men who also came to tea and sat balancing cups on their capacious laps even absent-mindedly fingered the "stuff" as they call it. They are, like Ned's Papa, in the Manchester textile business, and are expert in appraising silks and cottons. I felt like a specimen of the fashionable world. They don't seem to wear the material they weave, but send it "down South" which you would think was thousands of miles away. It is odd, but there seems less thought of rank up here and that pleases me. I can see where Ned got his easy manners. He is to return next week so I expect I shall not have the time again to write you such a long letter. Dear Sarah, do write and tell me of your doings and what you are going to decide about you know who. I can thoroughly recommend marriage. With love, Marianne.'

Maybe it was this letter, in conjunction with René's, or maybe it was that the long weeks that seemed more like years since Kate and Marianne had married had made Sarah rather panicky about her 'boredoms', but the next day she went out riding in the afternoon to think it all over. Would Mama and Papa be more pleased for her to accept René than to stay on at home and wait until perhaps (but it was a very uncertain perhaps) some other suitor turned up? Sarah rode a long way, the only sound that of the hoofs on the frosty turf and the trees bare against the pewter sky. When she thought she had gone far enough, she dismounted and tethered Star and sat on a wall and talked her thoughts aloud, as she had done since childhood. 'You see, Star, I am not much use here. What I really would like would be to work as a nurse or a groom or even as a race horse owner! But there is no war at present where I could nurse and I have no money with which to start racing stables. I should like to have children and to share my life with a man. And I have been asked. But I am not sure whether I should like living in France. Would there be some way of persuading René that I must spend part of my time in England? Would there be work for him here? What sort of work is he going to do in any case? Will Uncle Paul leave enough money for René not to have to work? He is supposed to have studied the law and then to have worked for his father, but he told me he did not like business and there is trouble at present in the vineyards. So it will have to be the law and I believe that is a very different thing in France from here. Would he expect a dowry? Would it be wrong to marry your almost first-cousin? Do I know him well enough? If only I could talk to Mama, but she is vague, and Papa will only say I must do what I feel will make me happy. Do I like René enough to marry him?' Here she added, 'Well, Marianne found she could like Ned well enough to marry him and I suppose Guy liked Kate well enough too. Would I be giving up all the dreams I used to have of running stables and marrying into the County and living near here with six children? That is what I'd like to do. But it may never happen, even if I wait ten years and in ten years I shall be twenty-nine, nearly thirty, and no one will wish to marry me.'

She buried her face in Star's mane for the mare had nuzzled up to Sarah at the sound of the familiar voice. What is she replying to me? thought Sarah. Is she frightened I should leave her if I went to France? 'But you are getting on now, old thing,' she said. 'And I should always be coming home. I should have to make René promise I could come home a lot. I wonder if he would mind. Does he just like me or does he think I am a good "parti"? I wonder if he has had many women. Frenchmen are always said to be very naughty. But he is half English and we have known each other well enough and he always liked me. I expect he thinks it is time he settled down.' The monologue did not decide anything, but made Sarah feel better.

She mounted Star again, feeling calmer with the clear factors separated now in her thoughts – money and residence and children – and the possible and the unlikely a little sorted out.

All the next week Sarah hunted hounds early in the morning and came in pink and breathless. In the afternoons she went into the library and sat by the fire letting her thoughts drift. She had often found that if she applied herself to sorting out the issues of a problem – but usually it was the illness of a horse or what to plant in her own plot of garden – then let her mind freewheel, the answer came up out of a deep well. By the end of the week she had remembered that she didn't like flirtatious men and that René was not exactly 'flirtatious' – at least he had never flirted with her! So that was a point in his favour. Then she had thought over the few young men in the district with handles to their names and decided that she did not like them, even if they were interested in her, which they did not seem to be. She dwelt on what she knew of René and tried to build up a picture of him, but it would not form itself very clearly. She thought he was serious about something, but what it was she did not know. She would have to find out.

She sounded out her parents about their wishes and guessed they both wanted her to marry. She let her mother talk at length about her half-sister Susan, René's mother, and led her father gently to talk of what could or could not be put into a marriage contract about the place of residence for future spouses.

'Papa, I want you to write to René and ask on what he intends to live,' she said, finally. 'I know it is bad form in England to discuss these things, but they would not think it so in France. Mama is not sure, but she thinks that René will be quite well-provided for by his Papa.' She blushed and Theo thought what a practical girl she was.

'You don't need to marry, Sally, if you would be happy to go on living here with us. You like it here, do you not?'

'Yes, you know I do, but I feel there is nothing I can do here that is not done by someone else – and Roger and Ralph will take over the management when they leave school, so I think it would be better for me to marry René,' she said. 'Except I have to talk to the Reverend Ferguson about a wedding with a Catholic. There is so much I must know, Papa, before I can reply to René's letter.'

When she had gone out Sir Theo was puzzled. If she wanted René Boissier she must have him and she must have an agreement that she spend some time every year in England. If the chap were keen enough he would not object surely. Sir Theo was lazy, but he gathered his wits together and wrote to his sister-in-law.

Sarah wondered whether she had been wrong to speak as she had to her father. What a business marriage was! Yet Ned had carried off Marianne with little fuss or prior arrangements. René should, by rights, have married Kate, she caught herself thinking. It was strange how they had married the men whom no one would have expected them to marry. And she? Had there ever been any other choice for her? René had proposed: René must have an answer. She must decide whether what she wanted above all was to be a good wife and mother. It was a not ignoble calling. She must be brave and confront her fate and learn to show tenderness to a man rather than a horse. She smiled at herself as she thought that. She must pull herself out of the slough of laziness into which she had fallen. Life was waiting for her.

In the event, Monsieur Boissier died a few weeks later at the beginning of February and René came over to Lake Hall Park with his mother. Sarah had relied upon her father to put the issues plainly and he did not let her down. Aunt Susan had much to say on the problems faced by English girls who married Frenchmen. René and Sarah went for long walks when they discussed religion, money and horses, but not necessarily in that order. René expressed himself delighted for her to spend some part of the year in England. A position had been found for him in a large firm of *avocats* and along with the investments left to him by his father (fortunately realised before the disastrous vine harvest of the previous two years) there would be more than enough for a decent life together. The religious problem melted away. René was not *croyant*, and his mother was, in any case, an Anglican. He had been brought up a Catholic, but had no objection to marrying in an English Protestant Church. Neither would he insist his children were brought up as Catholics, though he thought they could be baptised as such.

In all this he never once mentioned the word 'love', and neither did Sarah.

Sarah Gibbs made her jump into matrimony because she felt it was expected of her; because she feared to vegetate for ever in the country and because René asked her – and she had always had a soft spot for him. Any reservations she might have had were unvoiced except to herself, then and in future. Once she had made up her mind she resolved to reserve a little part of herself which could not be swallowed up by her husband or her new state, and to keep it to herself.

The wedding was to be in September to give time for all old M. Boissier's affairs to be attended to. They would honeymoon in Italy, return to Paris and spend their first Christmas in England. All was set fair. As for René, although he vaguely realised that Sarah was his moral superior, he did not look below the surface of the kind, if slightly eccentric, girl. He knew nothing of her unrequited feelings for Ned Mortimer, for example, nor that Sarah was the owner of a passionate sensuality, for Sarah only half understood this about herself. He looked forward instead to advising her on the dress and deportment necessary in Paris. It was quite chic nowadays to marry an English 'milady' – and a good rider to boot! Sir Theo had scrimped up a reasonable dowry for her and as for Sarah spending a portion of each year in England, this suited him quite well for it would give him the freedom to enjoy himself in his old ways. He did not intend that he would always accompany his wife to Lake Hall Park as the years went on.

Ned Mortimer looked at his baby daughter with an expression of tenderness and joy on his face. There was no doubt the baby looked like its mother. The hair would grow darker, but the eyes might perhaps stay that deep, dark blue which he so admired in his wife. Now that she was nearly three months old she did not seem so tiny and fragile. Not that Louisa had really been fragile, as might have been supposed for a child born a month or two early. It had, he thought, been the result of all the travelling and the unaccustomed changes of the first months of their marriage which had brought the baby early into the world. Though Marianne had not seemed tired or ill. Indeed she had given birth easily and was now a perfect mother, if perhaps excessively fond and anxious, even when she was now well and hearty again. She must not go through another birth too soon. She was only twenty and there was plenty of time for other children . . . He let the baby hold his large third finger in her tiny fist and as he stooped over the cradle and the sounds of late August came through the open window, he was unspeakably thankful that he was now married and a father, set to enjoy a peaceful and busy life with a wife of whom he was very fond and a child he knew he would love. Perhaps in another year or two they might have a son.

*

The couple stopped by a large, ochre-coloured church out of the October wind, and, 'Why, it's Sarah,' said an American voice. And the voice's partner turned and doffed his large, brown hat.

'René, old man! What brings you here?'

'Only our honeymoon!' replied his friend, laughing. 'It's become rather a protracted one.'

'They must come immediately to our apartment – it's in a Palace!' cried Kate, the first remark to Guy and the second to her two old friends.

Sarah was hanging back a little, wondering whether to comment upon the long months when Kate had not seen fit to write or communicate, but Kate said, 'Miles will be awake and you must see him!'

'Miles?'

'Why, yes, our baby – you didn't know we were Papa and Mama now!'

Anything less like Mother than this slim, sophisticated, cool woman, who seemed to be in charge of her husband, Sarah had never seen, but she tried not to gape and said, 'But I had no idea! Congratulations! You know Marianne too has a child?'

'Yes, we received a *faire-part*,' replied Guy. 'Ours is a few months old now – and looks just like Kate. We have a very good Italian girl as well as a housekeeper and Kate's maid. Servants are so cheap here, don't you find?'

'So we are quite free to enjoy ourselves too – and Miles is *such* a good baby.'

Sarah was impressed by Kate's household when they visited the couple one evening a few days later. Kate had spoken the truth. A Roman palace might not have seemed the best place for a nursery, but here Guy's astonishing practicality had asserted itself and a room with a view over the Villa Borghese had been adapted for little Miles and his nurse, a fresh-faced peasant girl. There were many other rooms, all furnished exquisitely. The baby was large and cheerful.

'Kate had a bad time of it – she didn't even realise she was expecting!' Guy confided to René. 'We had everything to sort out in a hurry for we hadn't imagined there would be an addition to the family.'

René raised his eyebrows and laughed. 'Oh, it hasn't changed our lives,' confided Guy. 'We have such a good nurse and Kate has made friends with lots of Americans as well as some of the diplomats here. The climate suits us – rather hot in summer, but then we shall go to Tuscany. Kate likes Florence.'

'You're not to come back home, then?' asked René. 'It would be chic to have you near us. I've bought an apartment near the Bois so Sarah can carry on with her riding – don't stay in Italy for ever, old chap.'

'That's for my wife to decide,' said Guy but in a tone of voice which said that he didn't mind. 'She has launched herself here, and found me lots of friends too. I don't mind telling you at first when we discovered we were going to have a child I was scared. But she said it wouldn't make any difference, and it hasn't. We're very happy.'

Sarah thought, too, that Kate seemed very contented. She had not changed at all; maternity had not given her that heavy brooding look some women acquire when they become mothers. The women were chatting in the nursery.

'The only thing I could not bear was if I were expected to feed him myself,' said Kate, with a grimace. 'I was determined to fit all my nicest clothes within a month of little Miles's arrival – and I did. Francesca is our wet nurse – Miles's cow we call her – and all I have to do is amuse him now and then. Who do you think he looks like?'

'Apart from the colouring, you, I think,' replied Sarah, bending over the bright-eyed boy, who smiled back at her. 'Or your Uncle Adolphe perhaps – his eyes are dark, aren't they?' She straightened up. 'I may as well tell you . . . I think I am expecting too – ' She blushed slightly.

'Oh, wouldn't that be swell for us all to meet with our babies – you will have hundreds, Sarah. And how is Marianne's little one?'

'I have not seen her yet, but they say that Ned is a doting father.'

'I'm sure he would be,' said Kate meditatively. 'I can tell you I was surprised when she married Ned. Of course, *we* decided to marry in such a rush and there was so much to arrange, that I scarcely took it in that she had married too. Is she happy?'

Sarah wondered why Kate was so interested, not having even sent a present (Marianne had said) although she, Marianne, had sent a silver mirror out of her own savings to Guy and Kate. 'Yes, it is all strange,' said Sarah. 'There we were, I thinking Ned would marry *you*, and – '

'Ned marry me!' cried Kate in astonishment.

'Well, he was sweet on you, wasn't he?' asked Sarah in her honest way. Perhaps she had gone too far.

But Kate replied, 'Oh, no – I don't think so. I don't think he likes me very much, you know.' She led the way back along the high-ceilinged passage to the *piccolo salotto*.

Their conversation was broken off as René came into the salon, followed by Guy and a servant with a tray of tall bottles in ice buckets.

'You must sip our Frascati,' said Kate. 'Isn't it just dandy to be together again? Quite like old times.'

Sarah stopped herself from saying that Ned and Marianne were missing to make it quite like old times, for she saw that Kate had firmly closed her mind to any reconsideration of the past. The evening passed pleasantly enough and she was allowed to nurse the son and heir and was complimented by Guy and told she would make a wonderful Rubens Madonna. Afterwards, on the way back to their hotel in a carriage pulled by a sleepy horse, she said to her husband, 'Why do you think Guy married Kate Mesure – did you ever find out?'

'Because she asked him,' replied René simply. 'And *mon ancien ami* likes to oblige.'

The three young women whose lives had up to the time of their marriages been superficially similar, were now launched upon very different existences. Kate, who in her Roman palace was invited to more and more of the most fashionable and glamorous of expatriate activities – balls, soirées and receptions. Sarah, whose husband had bought her a new horse, Cigar, as a wedding present, found she could not ride in the Bois as she had planned that year, for her pregnancy made her very sick. She, who looked to all intents and purposes the healthiest of the three and certainly the most physically active, had a difficult time before and during the birth of her first child in 1869. Olivia Penelope Susan took after both her father and mother in being long-limbed and fair with René's discreet air about her.

Marianne, who had been a nervous child, was also a nervous mother. She adored her baby girl and Louisa repaid that adoration by being a noisy and demonstrative baby who delighted Ned and her grandparents. She was to grow into a rather wilful little girl, but very sharp as they said in Westmorland. Marianne found time to read, once her household was established, and often took a book up on the nearer fells in the summer, glad to leave her domestic duties to the servants. Their Georgian house only half a mile from old Mr and Mrs Mortimer's mansion, was half hidden by woods and looked out over fields that in the spring were filled with daffodils. Marianne took Wordsworth along with Chateaubriand and Hugo and began to feel that she really belonged in her adopted country.

Sarah did not manage to feel that she belonged in Paris, but so long as the holidays were spent in England at Lake Hall Park she was content.

The women kept in touch with each other through Sarah, to whom both Marianne and Kate wrote, Marianne regularly, Kate irregularly. Marianne and Kate never corresponded, but knew of each other's doings through their mutual friend. There were naturally topics which none of them ever wrote about. Sarah could have told her friends of the passionately successful physical relationship she enjoyed at first with her husband, but one did not write of such things; Marianne could have confessed that, in spite of her great liking for her husband, he never excited her to loss of self-control: when Guy had kissed her hand on the balcony at the Warrenders' ball (it seemed hundreds of years ago), it had been more exciting than married love with Ned. Kate could have spoken of her indifference towards the physical side of her on the whole pleasant relations with Guy, but she preferred to exult in the painting Winterhalter did of her in the fourth year of her marriage when they returned to Paris after the terrible events of the Commune.

Sarah's second child, Valéry-Paul-Armand-Théodore, was born during these years and spent much of his babyhood at Lake Hall Park with his mother and grandparents.

Marianne heard the rumours of war but was spared the reality and was busy with her family. Louisa soon had a brother, then a sister, Hugh and Emma.

Kate returned to Paris with a new baby, Kitty, when the third Republic was announced. She had in the meantime exchanged her Roman palace for a villa in Tuscany, and found that social life in France was carried on in much the same way, American money remaining just as useful and welcome.

The fashion moved from voluminous crinolines to bulging bustles and Kate and Sarah and Marianne, still comparatively young women, accommodated themselves to its changing dictates as far as their physiques and purses allowed. Prices went up in the Seventies and even Kate found that the pursuit of pleasure had its cost. But she did not abandon her active social life one whit. Even if it glittered less, once the Imperial Court had gone from Paris, there were always other places – Monte Carlo, Baden and Vienna and her beloved Italy. She rarely visited England, but from time to time saw Sarah, who she thought was still full of *noblesse oblige*. René Boissier, it seemed to Kate, did exactly as he pleased – he had made a good decision when he married. Sarah reported to her that Marianne was very happy in her new provincial life, always busy and obviously a devoted Mama. Kate did not envy her *that* way of life.

Between the Acts
Christmas 1877

13

The two women stood stock still looking at each other for a moment before one of them moved. It was Marianne who forced herself to advance. She had seen a beautiful woman staring at her and had forgotten herself in this sudden realisation, before she had even recognised her. Here indeed were beauty and elegance. Was it an illusion? The figure, taller than she had remembered it, in a long watered silk dress – a great black triple bow at the back of the waist, tiny grey gloves at the end of ruched sleeves, a black . . . bonnet, she supposed she must call it, with a yellow rose perched in froths of lace slightly left of centre of the jet black upswept hair, the lace falling down the back of the bonnet like a miniature dark Niagara.

But when the figure spoke in a flat American voice and said; 'Why it's Marianne', her trance broke and she replied; 'Kate – why have you not taken your gloves and bonnet off?' What a stupid thing to say. She was not the hostess in any case and Kate had obviously come straight up to the gallery. Now why should she have done that?

'I thought I'd like to look round the place – Guy has gone for a walk round the lake,' said Kate.

Marianne drew together all the reserves of strength she had known she would need the minute Sarah had announced that Guy and his wife were to join the party and replied coolly; 'That doesn't sound like Guy – it *is* rather cold.' But how would she sustain this coolness when she actually saw him?

Kate took her arm and it seemed for a moment as though eleven years did not separate them from the days they had spent together during their season and that they were girls together again. Marianne cleared her throat. 'Perhaps Ned will do the honours,' she said, looking out of the window, 'since René is not yet arrived.

Sarah and her mother have been so busy organising everything. Where are your children? I must see them.'

'Oh, Sarah's two took them up to their quarters,' replied Kate. 'I had forgotten how beautiful England can be in winter,' she went on.

The afternoon light was fading now and the light powdering of snow gave to the park an eerie light and even entered the rooms of the Hall to bestow a sort of artificial dawn at four o'clock. 'You never want to live in England then?' Marianne asked as they made their way down to the drawing room. She stole another look at Kate as they descended the stairs and saw a healthily coloured cheek, arched eyebrows, a long nose, deeply-set eyes and a mouth shaped like that of a beauty from the Italian Renaissance. Long silver and pearl earrings swung from thin lobes and altogether Kate was a picture. 'You are looking excellently well' Marianne said, and felt rather foolish.

'Oh, I have a good dressmaker and hairdresser,' replied Kate, laughing. 'Don't you think it would be fun just for the three of us, Sarah and you and I, to take tea and leave the men outside and the children safely tucked away for one hour or two?' It was not like Kate to want female company. Marianne wondered what she wanted, then decided she was being uncharitable. What wrong had Kate ever done her? It was Kate's husband who had wronged her, if anyone had. Kate had known nothing about it, she was sure. 'Of course there will be scores of other guests staying over, I expect, and we have only three days,' Kate went on. 'Where will Sarah be?'

'In her little sitting-room, I expect,' said Marianne, leading the way.

'I am longing to see *your* children,' said Kate.

'And I yours,' said Marianne. 'Look, here are Valéry and Olivia.'

Sarah's two had escaped surveillance and were chasing each other up and down the long corridor at the bottom of the stairs.

'Their nursemaid will be looking for them, I suppose. I imagine English children are usually to be found in the nursery,' said Kate in answer.

'Oh, not Sarah's. Of course they are French, but Sarah is very free with them.'

For a moment they stood looking after the disappearing velvet knickerbockers and beribboned hair and listening to the high-spirited yells. Marianne opened the door and found Sarah sitting in front of the fire wrapping tiny packets in green paper. 'Thirty necklaces,' she sighed. 'For thirty little girls – I must be mad. Why, Kate, there you are'. I thought you'd gone up to your room.'

'No, I thought it would be nice and English to chat at the fireside and toast crumpets,' said Kate.

Marianne decided she was not being ironical.

'I'll ring for Brown, then. The servants usually bring the tea trolley about now – I wasn't sure if you took tea. Mama is in her room resting,' said Sarah.

'Not if it's any bother,' said Kate. She parked herself elegantly on a small sofa and sat surveying her two friends. 'I'm longing to see Miles and Catherine,' said Marianne.

'We call her Kitty,' said Kate.

'And you must talk French to Marianne's Louisa,' said Sarah to Kate. 'She is very lively,' she smiled. 'My two speak English better than I speak French, I'm afraid.'

'And when is René arriving?' The question seemed to throw Sarah into a little confusion which she averted by taking up a pair of tiny silver scissors to snip at the delicate fastening of one of the little parcels which were piled on a sofa next to her. 'Kitty will help you do the presents for the party,' said Kate without waiting for an answer.

'Oh, there's no need. They are supposed to be a surprise. I thought lead soldiers for the boys,' she said. Then, 'René will come over in the New Year – he has business, I'm afraid.'

'And Ned is busy, too,' added Marianne. 'But he wanted to be with the children at Christmas.'

'Of course Christmas is not such a thing on the Continent,' said Kate, perhaps to soften any implied criticism of Sarah's husband in Marianne's words.

'No, and he does stay quite a lot of the time down here with us – ever since Father died. I feel Mama wants so much for me to bring the children whenever I can.'

'I was sorry about your father,' said Kate. 'How are your brothers managing the estate?'

'Well, Ralph is like Father and prefers to bury himself in the country and Roger is still at Oxford.' She did not say that Ralph was lazy and Roger a bit of a dull dog.

'And how is Paris?' asked Marianne. 'Have things settled down now?' She had never elicited all the details of the Commune from Sarah who had fortunately got over to England before the worst excesses. It would be interesting to know how Kate had coped.

'We were still spending most of our time in Florence, although Guy was in Paris during the worst week. He doesn't say much about it now,' said Kate.

'I'd only been married two years and had Olivia with Valéry on the way. René insisted on staying with his mother once he'd seen us safely home,' said Sarah reminiscently.

'I believe René was with your husband when the Communards fired the Tuileries?' said Marianne to Kate.

'Guy was near Père Lachaise, *painting* if you please, or at least, with his sketch-book, that dreadful evening when they fought among the tombstones!'

'What a picture that must have made.'

'Yes, but he wasn't satisfied with his sketches. Threw them away and never finished the oils. I was furious I can tell you! Risking his life and all for nothing.'

'It was all dreadful,' shuddered Sarah. 'It doesn't seem possible, sitting here seven years later, that it happened at all.'

'England always softens everything – makes it rather *mauvais goût* to be actually fighting a civil war.' They were all silent for a time and then Marianne said, 'And here we are now with our children – thank God we are all safe, not that living in Westmorland is especially dangerous.'

'Speak of *petits diables*,' drawled Kate as the noise of many footsteps was suddenly heard and the door was opened by the maid with the tea trays and trolley which were almost overturned at the same time by a host of children.

'Careful,' said Sarah and rose and extricated the maid. The children were suddenly silent when they saw three mothers, then all seven of them made up their minds to file in. 'Where can Nanny and Maud have got to?' said Sarah, but without much interest. She threw a large silk shawl over the little parcels and said, 'Come along in and say how do you do to Aunt Marianne. And, Valéry and Olivia, here is Aunt Kate, Mrs Demaine.' One by one the children, now quiet, advanced. First Olivia and Valéry Boissier, both tall like their mother and father and with more than a look of Sarah's mother and father about them. Olivia, who was eight, curtsied to both Kate and Marianne and looked cheerful enough.

Valéry, a year younger, held back but consented to look out of the corner of his eye at Marianne.

'That's my Mama,' said Louisa Mortimer the oldest of the bunch, a dark, pale, thin girl in a pale blue dress with a drooping hem. She turned to Kate and said politely, 'Hello, I'm Louisa Mortimer. Will you speak French to me, and Italian? Mama says you live in Europe – I should so like to live in Europe.' Her sister Emma, a sturdy child of five, had been fighting her brother Hugh. He was a pleasant-looking, quiet boy who shook hands with Kate as he had been taught to do, saying, 'This is my little sister, Emma. She has left her nursemaid at home. Mama, she wants to go to the water closet.'

'Oh, dear,' murmured Marianne and rose to take her youngest child upstairs.

'Mama, this is Miles,' said Louisa disclosing behind her a tall boy with a look of Kate.

Marianne stared at him. Guy's son meeting Guy's daughter. Oh, how she had dreaded this moment, and how she still dreaded meeting his father.

'And this is Kitty,' said Kate as the smallest of the seven approached and went up to Sarah and stroked her dress.

'*We* all know each other,' said Louisa. '*I* am the eldest – Miles is nine too, like me though, and he says I can play skittles with him in the cellar.'

'Now Louisa, just calm down,' said Marianne, scooping up her Emma.

'And there will be a big, big party,' said Olivia. 'My Mama is going to give you a big, big party with hundreds and hundreds of girls and boys . . .' She lisped a little, had rather large front teeth.

The children came up to the fire in a circle as Marianne went out with Emma. Louisa could not take her eyes off Kate whom she thought uncommonly elegant, not at all like Mama or Aunt Sarah.

'You must ask your maid to hem up your dress,' said Kate to her.

'Oh, dear – is it coming down again?'

'*My* hems were always descending,' said Sarah. 'Now all of you, a piece of the cake. And, Mama, I'll clear the children out if they bother you.' This as Penelope Gibbs came in, stouter now and greyer.

'All together at last,' she said sentimentally. 'You know, Sarah has been planning this for an age. Isn't it lovely? Just like old times, except now you have all these children!' She gestured vaguely around and Valéry chose the moment to say, 'Emma is with her Mama. She had to go to the water closet.'

'Hush, dear,' said Sarah while Kate smiled.

'Here come Maud and Nurse. Now, Nurse, if you would be so kind as to arrange these young people on these small stools and cushions, I think you might then go and finish dressing the tree,' said Lady Penelope.

The children submitted to being arranged and handed plates and sat steadily munching. Marianne returned with her problematic daughter and poured tea for the grown-ups.

'Isn't this fun?' said Miles. 'We never have tea in Italy or Paris – oh, I do like England.'

Louisa handed him a piece of cake and contemplated him. He was the most handsome boy she had ever seen, just as his Mama was the most beautiful lady she had ever seen. Aunt Sarah was kind and nice, but they had met her before. She had even been to stay with them at Langthwaite with Valéry and Olivia who were rather shy then, but seemed to have improved on home ground.

'We ought to have all the children photographed,' said Lady Gibbs. 'I've sent for Mr Beningborough from Oxford to discuss prices. What about the day after tomorrow? Too much to do tomorrow with the party. My, my, I think we may be in for more snow.'

Marianne heard voices then and wondered when the ordeal of meeting Guy would be over. She could not avoid it for ever. Indeed, one of the reasons for accepting the invitation had been to trust herself to see whether she must spend the rest of her life fearing to come across him by accident. Originally Sarah had invited her and Ned and their children, but then it turned out that René insisted on Guy and his family coming too. This was nothing though to the fear that her daughter, *his* daughter, Louisa, would meet his son, her half-brother Miles, and in her headlong way become attached to him. It was absurd. They were only children. But Louisa had a very pronounced 'temperament'. She was like Marianne but less of a physical coward. Ned doted on her as on all his children. Hugh was a smaller edition of his father and Emma the picture of Ned's Mama, and all three children had enough of

Marianne in them for there not to be any doubt they were full brothers and sisters. The time had long passed when Marianne had thought to confess herself to Sarah. It would be wrong to make Sarah bear a burden for her, wrong and self-indulgent. Sarah had enough to worry about. René, Marianne gathered, went astray – with fashionable cocottes and even one of their own friends' daughters, a girl of only nineteen. Poor Sarah. He did not mind then that she spent a good deal of her time in England. Her mother too must have guessed that all was not well between the two of them, though Sarah had never uttered any criticism of her husband.

Lady Gibbs looked into the fire and thought how cosy it was. Just the mothers and the children – well, men were not expected to come in for tea, it was almost a nursery in her daughter's sitting-room. Sarah was a fine woman going to all this trouble to see that the village children joined in the party tomorrow and doing most of the work herself. After the Christmas celebrations she would be off riding, her mother thought, still happier on a horse than anywhere else. Her children did not really take after her. Dear little Valéry was the image of her own late husband, good-natured, but cleverer than Theo had been, and Olivia was a little French Miss, even though she was fair and tall. Penelope Gibbs cast her face up to look at the other children. Louisa reminded her of someone, she couldn't think who – she had a look of her mother of course and was a rather talkative child at present, whilst her little sister, Emma, was quiet and placid and her brother Hugh an English squire in the making with a ruddy complexion. Kate's children were surprisingly nice too, though why that should be such a surprise she was not sure. Guy was, by all accounts, a good husband and father – enjoyed gadding about, but spoilt his children, his daughter anyway, Sarah had said. The boy, Miles, was a handsome lad – it must be the American blood. He was tall and had that broad American mouth and a frank, free air.

'A photograph,' she said again aloud, thinking, Olivia and Valéry and Miles and Kitty and Louisa and Emma and Hugh – seven children of the seventies with all their lives to enjoy before them if measles or scarlatina or other misadventures could be avoided. And all from matches made at Lake Hall Park!

Ned often thought of his wife, even when she was somewhere near. He was an uxorious man and rather sentimental. He remembered her the summer of Louisa's birth in the garden at High Pines lying in the hammock on a day of cloudless blue, and he remembered that he had felt happier then than ever before in his life. There was always the worry over business and markets and exports, but when he thought of his wife and of his children, he felt warm in the region of his heart. He had been so right to marry Marianne and he was still surprised that she had accepted him so readily. He knew she had been glad to get away from her parents, that ever-striving couple – but after all they had produced Marianne. Would *his* son-in-law say one day, 'Old Mortimer – an odd cove, but after all he produced Louisa'? Louisa was the cleverest of his children – took after her mother. Emma was the prettiest in his opinion and the boy, Hugh, the most sensible. But that summer – his thoughts went back to it as he walked round the park, dark now in the crunchy snow, with Demaine, who was rather quiet when you got him alone – that summer was a 'painting' to him. Yes, that was what it was; like a scene that had been painted on his inner eye so that whenever he was depressed he had only to recall the garden and the pale mauve crinoline she had been wearing and the yellow sunshade and the roses in her lap tumbled in with the lace, and her thin, brown hands, and the smell of those pink roses – ah, that was happiness. That summer, before the trouble in France, before the hard work and worry of the other two pregnancies, Marianne had been a flower herself, and the little baby, already brown and smiling, his Louisa, and the smell of the air and the sound of the lake waters below and the clouds reflected in the lake – a paradise. Some fellow had just painted a real picture he had seen in Paris when he was there on business and it had brought it all back. Marianne was always talking about Art and Beauty, but that was a lovely picture – women in a garden. They would never grow old as he and Marianne would. Why should he think of growing old? He was only thirty-nine and tried to keep healthy. A bit overweight, he supposed, looking covertly at Guy Demaine's trim figure in its Parisian galoshes and cape. 'Time to go in?' he asked. Never had much to do with Demaine, did not envy him his wife. But the chap was all right. Seen quite a lot of the Commune troubles too, about which he had been talking.

They went in, Guy yawning a little. He had hoped the cold air would wake him up for dinner, which the English in the country had so early. Pity René wasn't there, but still, they would be moving on soon to London to stay with Uncle Adolphe, and then back to the social whirl on the Continent. He looked at himself in a long pier-glass in the drawing-room.

His hair was parted fashionably in the middle and he wore a small, crisp moustache. He was older and heavier than he had been ten years before in this very drawing-room, but so were they all. His children were being ushered upstairs to wash and then to go bed, along with the Mortimer and Boissier sprigs. Before he went up with the others, his son Miles told him that no one would be allowed to see the tree in the long gallery till the next afternoon when it would be unveiled by Lady Gibbs. Other children would arrive for Sarah's party. Oh well, the day after that would be Christmas Day and he supposed he could avoid a visit to the church across the park. There would be the usual English meals and presents, and, he hoped, some of the best claret old Theo had once laid down.

Their *bonne*, who did not speak English very well, came down again and seemed rather agitated. He was used to dealing with Kate's servants – she left most of that to him, so he told her to find Mrs Mortimer's Polly, who would show her where to go and what to do. 'I expect there will be a nice English dinner for you in Servants' Hall,' he said and she gave a grimace, having much the same opinion of English food as she knew he did. When she had gone he yawned again. There was no sign of his wife. He had better go and dress for dinner. England always made him feel sleepy.

Marianne was in a gut-turning fever of dread over the time when she would have to meet Guy Demaine at dinner. She had successfully avoided it so far by concentrating on her children and talking to her maid, Polly Braithwaite, who rattled away as usual treating her like an equal. For once Marianne did not feel obliged to cut Polly's ruminations short as it postponed the moment when she would have to descend. Polly had already expatiated upon the holes she had seen in the curtains and the scratches on the furniture and the fact that beer was served 'even to the lasses' in Servants' Hall, and Marianne had listened when she really should have silenced her. 'Miss Louisa's too excited,' Polly went on. 'If she's not careful she'll be sick before the party.' Then Marie, Kate's *bonne*, who had put her charges to bed, came looking for Polly and she had sent them off to the servants' sitting-room to carry on with the parcelling, Polly wondering if the maids would get presents tomorrow too.

Ned was ready in his smart dinner tails.

'You go and talk to the others,' said Marianne. 'I'll be down soon.'

When she got to their room she took some deep breaths and pinched her cheeks and tried to calm herself. Would Guy say anything to her to show that he remembered? Surely not? He had probably forgotten. How was she to talk to him? She tried to imagine what Sarah or Kate would have done in such a situation, but it was unimaginable. She could not be her natural self. Perhaps she could plead a headache? No, it would only postpone the dreadful moment.

She took her fan and her handkerchief and her evening bag and listened at the door to see if Sarah were yet down. If not she would follow her. But there was no sign of Sarah. She would rely upon her husband to have gone straight up to Guy. That way Ned would be with her when she met him again.

She went down. They were all in a little ante-room, waiting for dinner to be served, a room she did not remember. What would she have felt if she had never had that afternoon with Guy Demaine or if there had been no result of it? She would have been pleased to see him, she supposed, and could have looked lightly back upon that lovelorn young woman. Might even have wanted him to remember their flirtation and to bask in the feeling that he had once found her attractive. But now? She had once thought herself a good actress. She had better put her talents to the test. She took a deep breath and went in.

No reaction. None at all. Guy bowed civilly to her and then turned back to Ned with whom he seemed to be having a conversation about pictures. She went up to Sarah after inclining her head graciously to Guy and sank down next to her with a feeling of great relief. Now it would be all right. He had greeted her politely, but without a flicker of real recognition. Perhaps he *had* forgotten? In spite of herself she could not help feeling a little hurt – and amazed. Their adventure might never have been.

She saw him being most attentive to Kate when Kate finally appeared, and then Ned came up to her and gave her his arm for dinner. Kate was ordering Guy around quite a lot, Marianne noticed, but he did not seem to mind. At dinner Kate and Ned were next to each other with herself opposite, seated next to a neighbour imported to make up for the missing René, so in the intervals of listening to a shooting story of interminable length, she was able to see that Kate still seemed rather frosty with Ned. Poor Ned. He caught his wife looking at him once or twice and smiled. Dear Ned.

I wonder if she's forgotten how she once used to go riding with my husband, probably to make Guy jealous, Marianne found herself thinking rather uncharitably. It was all water under the bridge now. They had chosen their lives and might hardly ever meet again unless Sarah made her children's party an annual affair. She felt so relieved that this first meeting with Guy was over, although they had not yet spoken together, that she turned animatedly to the shooting gentleman and told him a hunting joke an old farmer in Westmorland had once told her. She had never seen its point, but the shooting gentleman obviously did, for he broke out in great guffaws and Marianne felt she was a success.

Kate was talking about racing, about the old winners of the sixties, Hermit and Blue Gown, and Sarah was remembering the horses in detail. Marianne wished there were someone to talk to about Mr Ruskin or Mr Arnold whose latest essays she had read with such enjoyment. Ned was not really interested in literature, kind, attentive and affectionate husband though he was. Sometimes at home she would talk to Mr Carmichael who had been Ned's best man, but he didn't have so much time nowadays, being so busy patenting some new invention to do with spinning mills. Uncle Lechmere would have graced this table and she hardly ever saw him either. The children, in spite of a lot of help, seemed to take up most of her time. She even sometimes neglected her reading. When she worried about the children nothing else could fit into her head. It was pleasant to be under someone else's roof and to know that if they were taken ill a doctor would immediately be ordered by Sarah to come round and she need not feel the awful responsibility. For if Ned was a good father, he was also often away. There had been that terrible winter two years ago when Emma was three and they had all had whooping cough and she had thought Louisa would die and she was to be punished for her past sin. How had she survived it? All these anxieties made one feel less young, which must show on her face, though she had been complimented by Kate – Kate! – for her good complexion that afternoon. She knew she was not bad looking, but compared with Kate, why, every woman would look provincial and dull. Kate had kept her figure, which she had not and Sarah even less. They both ate too much, that was the trouble. Not that she was fat, but she would never regain the slim waist of her nineteen years. Still, she would not have it different. She had produced three fine children and Ned appeared a contented man, except for the occasional days when he would seem lost in a brown study. But she was not worried about her husband at present. It was Guy whose face she wanted surreptitiously to study, without giving herself away.

Marianne saw she would never have anything to fear from him in future. He seemed and was a stranger, though he was the father of her eldest child, sleeping (soundly she hoped) in the children's wing. She found herself thinking, It is Ned who is Louisa's real father, Ned who has loved her, cared for her. Guy is nothing but a handsome stranger, dangerous only because he has another child, that nice boy Miles, who must at all costs be prevented from seeing too much of Louisa in future. The very thought made her heart beat. But why worry about what might happen in some distant future? She must concentrate on the here and now. Guy had clearly banished any remembrance of their passionate afternoon. It was she who was the fool, she who would bear the burden to the end of her life. It was Ned upon whom she must concentrate, her husband who had not deserved her, whom she had certainly not deserved.

She ate her pudding, half-listening to her hunting gentleman and said to herself, Until Louisa grows up, I shall pretend it never happened. It has not been too terrible after all to see Guy again and should be less terrible in future now that I have done it. She returned her attention to Kate and Sarah. Kate was looking even more attractive than in the afternoon, her hair now in a complicated, high, upswept style which Marianne had seen only in the latest fashion pages which arrived rather late in Westmorland. A long, wavy tress fell from the back and curled voluptuously on Kate's lacy collar and the crown of her head was arranged in three enormous ribbon-shaped pads with a jewelled comb holding up the whole structure. It must take hours to do and yet Kate had brought no personal maid with her. She still affected the American manners – help with the children yes, and probably a French cook, but she was independent as far as her own toilette was concerned. She was so gifted at making the best of herself. Marianne sighed. There was also quite a lot to start with, which long earrings and fashionable looks could not transform into beauty unless a substratum existed in the first place.

Now they were talking about Florence and how Kate had met Robert Browning, the lonely widower lionised by all the American ladies. 'It should have been you to meet him,' Kate said kindly to Marianne. 'I expect you read his poetry – you were always a one for verse.'

'Indeed, I should love to have met Browning,' said Marianne, crumbling some bread between her fingers. What a life Kate must have. 'But he is sometimes too clever for me,' she added. 'I am more at home with the old "Lake Poets". When you read Wordsworth not far from where he wrote his poetry, you understand it better.'

'And what about Ruskin?' asked Kate, who seemed *au fait* with the current English literary lions.

'Oh,' interjected Sarah, 'I read *Sesame* and *Lilies*. I have met English ladies in Paris for whom it is a sort of Bible – never took to it myself though.'

Marianne became animated. 'Uncle Lechmere and I have many discussions when he comes to stay with us about *Unto This Last* – I don't suppose you have read that?' She turned to Kate who smiled. Guy looked bored and Ned tried to follow his wife's remark with a 'He doesn't like modern times, old Ruskin, eh?'

'No, he quarrels with the great heroes of Uncle Lechmere – Mill especially. Uncle Lechmere agrees with Mill and so ought I to, I think, for I do believe Mr Mill is right.

'Mr Mill is all that John Ruskin abhors,' said Ned. 'He deplores the scientific spirit of the age, especially that found in Mr Mill's writings. For myself I am for female suffrage.'

Sarah gaped rather at Ned and Lady Gibbs cleared her throat. Tubby's son was certainly unusual. For the first time she wondered whether perhaps her Sarah was better off in Paris with a philandering husband than one who was so clearly a Radical. Marianne sensed this and thought she had better mollify her hostess. 'I am torn,' said Marianne. 'I agree with Ned about those things, but I also feel there is more to life than suffrage – nature and art and all our questions . . .' she ended rather lamely.

Ned caught her eye and smiled, knowing her ploy.

Lady Gibbs now looked vaguely approving and then Kate changed the subject quite cleverly saying, 'So many things happened before we girls were even out and now I wonder why we did not heed them.'

'What sort of things?' asked Sarah, who admired both Kate and Marianne's knowledge of the world that lay beyond the enclosed one of home and husband and children.

'Well, I never talked to you about our civil war did I? Or President Lincoln's assassination – yet Uncle and Aunt were in Washington at the time!'

'We were too busy thinking about ourselves,' offered Marianne, hoping the talk would not pass to men or love. But Kate turned it round again.

'You often spoke of the conditions of the poor – both of you,' continued Kate, turning to both her old friends. 'Yet it was not done to discuss them.'

Lady Gibbs looked as though it was still not a subject for dinner-table conversation, but Sarah's friends were both clever women, she knew. Her daughter said, 'We even have the new Salvation Army girls in Paris sometimes. I'm sorry for them. Somehow they don't seem to belong there, though I'm sure there is just as much vice and poverty in France as here.'

'Oh, come, come,' said Guy. 'London is a sink of iniquity. We don't need Protestant' – he was going to say virgins but changed his mind – 'Protestant nuns at home. I haven't noticed my wife taking much interest in social conditions.' Kate was clearly annoyed, but said, 'No, I expect it would be hypocritical.'

'The Church . . .' murmured Lady Gibbs.

Ned said, 'It is a change from discussing Mrs Beeton's receipts and the problem of servants – not that my wife seems to find running a household too much of a problem.'

Now it was Marianne's turn to feel annoyed. What did Ned know of her attempts to organise the servants and to get Cook to prepare something other than the plain fare of the country? Men were all the same. Guy and Ned might in their different ways want their wives to have opinions, but they did not perhaps envisage a fundamental change in society. Marianne was still pondering the difficulty she had in believing in both Mr Ruskin and Mr Mill and she wondered what Kate really believed. You never knew with Americans.

'It's a scientific age,' said Ned. 'We have to move with the times. All those sixties events – your war (he barely acknowledged Kate) and our social problems – they have to be dealt with before we can improve our lives. Of course, I still feel that only by carrying on successful trading and making profits can we improve the lot of the masses.'

Sarah, who had been silent, put in her oar, turning to Ned. 'Were you in London in '71? It was just after Valéry was born and René was still in France and I was lodging with Aunt Leverton – did you know that there was almost a revolution that winter? I went for a walk in the snow in Hyde Park with my *bonne* and Olivia in her little perambulator and the Serpentine had frozen over. People were skating there. I should have loved to skate too, but had no ice skates. And then we noticed that when the people wanted to get off the ice – there were lots of sparks and their ladies – there were lower-class people (she blushed), louts really, who demanded a toll. They wouldn't let the girls or their partners back into the park unless they paid. It was quite frightening. No one said anything about it that I know of – I can't remember reading about it in *The Times* – but apparently it had happened before, oh, before we went to Madame Duplessis. There were riots, you know, in the West End – '

'I suppose John Stuart Mill wants to reform the louts,' said Guy with a shrug.

'I don't know, but it must have been frightening. One of Mama's friends, you know, Lady Dove was there. I remember her telling Mama – do *you* remember, Mama?'

'No, I can't say I do,' replied Lady Gibbs. 'I do remember Theo saying that if they brought in the suffrage bill in '67 there would be trouble.'

'But these sorts of louts don't *have* votes,' said Marianne, then stopped. She thought, No, they are on a par with all women. Women and low-class louts. They have no vote.

'We live in changing times, even up in Westmorland,' said Ned.

'Well, we all know which side our bread is buttered on,' said Kate with a laugh. 'I changed my mind about taking an interest as Guy calls it. We might have to beat revolution at its own game.'

'England will never have a revolution,' said Ned quietly. 'Though I couldn't speak for Italy or even France . . .'

Sarah shuddered. 'It was too terrible,' she said. 'René was in danger, as you were too, Guy, but it's all calm and peaceful now, Paris – '

'Yes, Paris is once more the capital of civilised life,' sighed Kate. 'Guy is pleased. It means we can settle there when he has had enough of Italy.'

'Oh, I shall never have enough of Italy,' said Guy. 'It's you, Kate, who prefers Paris. I must say, now with Bernhardt at the Comédie Française and the new painters, we are set for an epoch of brilliance in France.' Not like frosty old England, he thought swallowing the last of his rich pudding with its white sauce. He cracked a nut and took another glass of Barsac when the butler came round.

'Talking of skating,' offered Lady Gibbs after a silence and when the women were expecting her to rise so that they could follow her to the drawing-room, leaving the men to their cigars, 'I thought the new "Rinkomania" – is that what they call it? – might do for the ground floor corridor after Christmas. We must give the children plenty to do and I've had Jenkins buy some pairs of roller skates in Oxford. Do you think they'd like that?'

'You mean turn the corridor into a skating rink, Mama?' exclaimed Sarah.

'Well, I did think of the gallery, but it would spoil the floor, so I wondered about the west corridor. It's very long and they might like to learn to skate along it!'

'A lovely idea,' said Marianne. 'Better than ice skating – I always think that so dangerous.'

'It might be frozen over still after Christmas,' said Ned looking through the window. 'I could take Hugh out there if we had any ice skates – but no, I think your suggestion of roller skates is most imaginative. If one skates all the others will wish to do so.'

'You are very kind, Mama,' said Sarah. 'Always thinking of pleasing the children – and I wouldn't mind a little "Rinkomania" myself!'

They got up then and left Guy and Ned and the hunting man, who had been silent, to think up some other topic of mutual interest.

As they were tidying their hair, Marianne said, 'It's nice, just the three of us with the men – and I hope René soon. We couldn't talk properly if your Mama had a large party.'

'No, she prefers small parties now, and, of course, since Papa died she has had to make more economies. Lake Hall Park won't see many large house parties in future – that's why I thought of a children's party.' Marianne remembered the holes in the curtains.

'Children in England are spoilt,' said Kate. 'And in the States too – on the Continent they expect children to be seen and not heard. Just think of all the nice things children can do here – '

'Not to mention the books they can read,' said Marianne. 'Louisa is quite infatuated with Alice. I must say I like *Through the Looking Glass* better than *Wonderland*.'

'Do you still play your piano?' asked Kate as she stared at herself in the glass as though she too were trying to look through it to find another self, Marianne thought.

'Not much – I prefer to listen. There are some splendid concerts in Manchester,' replied Marianne. 'I can't go very often, but Ned takes me when he has an evening off from his labours.'

'Guy prefers opera and I must say I like to make an occasion out of my music,' said Kate.

'You can't avoid culture in Paris,' sighed Sarah and the other two laughed. 'I mean when Victor Hugo returned after the Commune it was as though they were fêting a great general – as though he had just won a battle. They all venerate him.'

'So they should,' said Marianne. 'It is so nice to find you both so caught up in the world. So many of my female acquaintances in the North have no time for anything but children and home.'

'Oh, I am not very educated,' laughed Sarah.

'What could you expect from Madame Duplessis?' sighed Kate.

'I'm determined my girls shall have a better education than we did,' said Marianne earnestly. 'I think Louisa is quite clever, you know, and there are the new day schools for girls popping up now, though I don't know of any near us.'

'Ever the blue-stocking,' said Kate, but not unkindly.

'Hugh will go, I suppose, to Harrow,' said Marianne. 'What about Miles and Valéry?'

'Guy is thinking of sending Miles to America,' said Kate. 'Largely at my insistence, of course.'

'René wants Valéry to be a Frenchman, not an Englishman,' said Sarah regretfully.

No one said anything further about René's wishes and Sarah did not pursue the subject.

'It would be nice if your Kitty and my Olivia could go to the same school, or will you just have a governess?' she asked.

'Convent, I suppose. Guy is a great believer in convents,' replied Kate enigmatically.

'So long as they have books,' sighed Marianne.

'The younger set may not care for books,' said Kate. 'Though Miles *is* a bookworm.'

The men rejoined them, but the women went on talking about their children in the manner of all young women with offspring. Marianne forgot that her Louisa was the daughter of Kate's husband, who lay elegantly on a *chaise-longue* in the small drawing-room looking at them both with the enigmatic smile they knew of old. Tomorrow would bring the party, with all the games and amusements to prepare, and there would be no more time for such talk.

Sarah looked out of the window at the vast expanse of parkland shining in the moonlight and wished that René were there and that she did not feel so lonely. Lonely in her own beloved Lake Hall Park with a new horse to ride and her Mama to consult and console and her dear friends around her! But lonely she did feel and missed her husband who she was sure would charm and enchant both her women friends, if not Ned.

But the next day brought him unexpectedly. He had said he would be away till the New Year, but something must have gone awry with his arrangements for his two children ran out in the snow when they saw him alighting from his carriage the next morning. All was confusion, with the Long Gallery commissioned for the party and forbidden as yet to the children. Marianne and the servants and Sarah were there dressing the Christmas tree when René looked in. The children had been shooed away and taken down to the kitchens by Nurse Hopkins. There they spent the rest of the morning, with the excitement mounting and speculation rife.

'I shall not know what to say to the village children,' said Louisa.

'*They* will not know what to say to *you*,' said Miles, regarding her with amusement.

He was used to a fairly chaotic life between his homes in Italy and Paris and took every new person in his stride.

'They will like the presents, I suppose,' said Hugh.

'Our Mama is preparing amusements for them,' said Emma. She had heard the others use the word and thought it probably meant something to eat.

'Let's go out,' suggested Louisa. 'Can we go out, Mrs Hopkins? I want to slide in the park.'

'I was told to keep you in,' grumbled Hopkins. 'But I dare say they won't mind if Lady Gibbs doesn't.'

Miles and Louisa crammed their woolly bonnets on and their galoshes and rushed out into the park from the servants' door. The others followed more gingerly. Once arrived at the front of the house they tore down to the lake over a path cleared by the gardener. All around was the whiteness of newly fallen snow and they set out to make a snowman without more ado. The younger children were deputed to ask the gardeners for a spade and Louisa and Miles worked away to make a passable figure. The sun began to shine and the snow glittered. Louisa felt excited, healthy, and enjoyed the fact that Miles appeared to accept her as an equal. So many boys despised girls and thought they could do nothing but sit and sew and nurse their dolls. But Miles was content to apportion the labour when the younger children came up with small spades.

'We need a pipe,' she said. 'Papa has one, I think.'

'*My* Papa only smokes cigars,' said Miles. 'But he might have an old scarf – or we could ask Aunt Sarah.'

'She's nice, isn't she, Aunt Sarah?' said Louisa.

'Don't pass personal remarks,' said her brother Hugh who was listening to them and cross because he had been deputed to stagger around with spadesful of snow, but not allowed to do the interesting part of patting the figure into shape.

'Well, she *is*,' said Louisa. She also would have liked to say to Miles that she admired Aunt Kate, but it was too personal to talk about someone's mother.

'When shall we have the Christmas tree?' asked Emma, who felt cold and wanted to go in, but did not want to miss anything.

'Mama is dressing it now,' said Olivia. She had been running down by the lake with her brother, so had not heard the 'personal remark'.

'Will the village children all have presents?' asked Emma.

'Yes, my mother has been wrapping them up for everyone,' said

Valéry. 'My father says she is very charitable.' He pronounced it in the French way.

'Do you speak French at home?' asked Louisa, patting the snowman's face into shape.

'With Papa we do sometimes, but he speaks English too. Mama likes English best and we stay a lot down here with Grandmama.'

'Have you a grandmama?' asked little Kitty, who was holding Olivia's hand inside her furry muff and watching her brother.

'Me?' asked Louisa. 'Of course. I have two – one at home in Langthwaite and one in Worcestershire, but she used to live near here when Mama came to stay with Aunt Sarah.'

'I haven't any grandpapa nor any grandmama but one,' said Kitty wonderingly. 'Mama's Mama and Papa are deaded and Papa's too, except for Mamie.'

'Who is Mamie?' asked Emma of this small, composed child who spoke a rather peculiar English.

'Papa's *maman*, of course, don't you know French?'

'No,' said Emma.

Louisa thought Kitty was rather spoilt and precocious, but Emma *was* very persistent.

'Look,' said Miles. 'I think it's finished.' He was panting with the final exertions. 'Let's go in for a pipe and a scarf.'

'I'm hungry,' said Emma.

'Perhaps we can have some *chocolat*,' suggested Olivia. The boys, Valéry and Hugh, were now dashing up and down the path and then stopping to gather snowballs.

'Don't touch our snowman with your snowballs,' screamed Louisa.

'It's *all* our snowman,' replied her brother Hugh, aiming a snowball at her which caught her on the chin.

'Right,' she said and scooped up an even larger one and began to chase him. He ran, turning and putting out his tongue at intervals as she flailed along and missed his retreating back. Miles and the others came up and then there was a general fight, the snowman forgotten for a moment. Mrs. Hopkins could be seen waving at the servants' door, not wishing to risk her shoes in the snow. 'All in,' she shouted.

They trooped into the back kitchens where hot drinks were awaiting them, all falling over each other to explain their morning. Polly helped divest the smaller children of layers and layers of packing. 'I wish I could snowball,' she thought and Miles echoed her thought, saying, 'We ought to get the grown-ups, and the servants,' he added kindly, 'to have a big snowball fight, I mean.'

'They won't,' said Louisa. 'But Polly will, won't you, Polly?'

Polly laughed. 'Too busy, Miss, with all them games your Mum's organising. You're to have a light lunch down with us, all of you, and then get dressed for three o'clock. Master's orders.'

'I want my Papa,' said Kitty. 'He said he'd play with me.'

'My Papa will play with you,' said Valéry. 'They'll probably go skating – the Frenchmen, you know.'

It was the way he said it – 'the Frenchmen', as though his father and Miles's Papa were two aliens imported into England, not like her own Papa. Louisa didn't think of her Papa as 'an Englishman'. Perhaps the children felt strange always travelling about – Aunt Kate's children too. Perhaps there was something to be said for being settled, and she thought for a moment of home up in the North and how the lake, far bigger than this tiddly one, would look.

'Penny for your thoughts,' said Miles, as she sat sipping her hot chocolate.

'Yours first,' she riposted quickly.

He considered. 'I was just thinking about nothing,' he said.

'How can you think about nothing?' asked Hugh. 'You have to think about *something*.'

'You could just think of the word nothing,' said Olivia, listening. They were such a clever lot these children of Mama's friends, she thought, and talked about such funny things.

'Yes, you could,' said Miles kindly. 'But I wasn't – I was really thinking of nothing.'

'But if you were thinking of nothing, you couldn't be thinking,' said Louisa after consideration.

'No, perhaps not,' he admitted. 'It was just a blank.'

'What do you usually think about?' asked Louisa in a low voice to him to exclude the others. Miles and she would be ten next year and she had quite decided she would marry him when she grew up. He was the nicest boy she had ever known. The village boys at home were quite friendly, but tended to be tongue-tied or else rough. She was not supposed to talk to them, she knew.

'Oh, I don't know,' he said. 'Perhaps about fishing, we fished in a lake in Italy, or about Robinson Crusoe – you know – '

Louisa, who had not yet read *Robinson Crusoe* was silent for a moment. She regarded her feet. The bottom of her pantaloons was wet with melted snow. Polly had brought down her flat black shoes, but her feet still felt cold.

'What were *you* thinking of?' he persisted.

'Oh, I was thinking my feet were cold,' she replied.

'My feet are cold, too,' piped up Kitty.

The *bonne* came in then to take the younger ones away for a rest preparatory to the afternoon's entertainments.

'I dare you to look at the tree – before the others,' whispered Louisa to Miles. 'We could go into the gallery before luncheon – while the grown-ups are eating their lunch – go on.'

'All right,' said Miles reluctantly.

René was always bored after a time at Lake Hall Park. But he came to please Sarah and because if he did this duty of staying in England from time to time he saved money and also earned freedom to do what he wanted in Paris. He continued his law practice but was not consumed with interest in the work.

Again it was a duty. A married man had to support his family and he was lucky that Sarah did not have expensive tastes. He had been surprised by the force of her passions when they were first married. The very fact of her sensual dependence upon him annoyed him and quite turned him away from the marital bed. Not of course that he would have wished her to stray, as so many Frenchwomen of their circle strayed from the connubial couch, but he felt slightly ashamed that she was not all in all to him as he was – or seemed to be – to her. He had married a good woman and found, as he had always known he would, that he still wanted other women who were not quite so good. Variety was his spice. He did not take up with *filles de joie* or even little *midinettes*, as Guy used to do; he preferred more exotic *demi-mondaines*. The pleasure of the chase was also strong. He knew his wife was aware of some of what was going on, though she never said, and her very pain made him feel a guilt he might not have felt with a shallower woman. Yet he could not stop himself. Sarah should be only a capable, kind wife and mother: that was her role. He was prepared to be fond of her, but he could not bear to be so passionately needed. A suitable subdued ardour was occasionally *de rigueur*, but not abandon, allied with a romantic desire for a soul mate! He just could not cope with that and had not found a way of dealing with it that satisfied her. There was the excuse of her pregnancy after the first few months, though her feelings for him did not seem to wane. And then there was the rather difficult birth of Olivia, which did take much out of her. He had been patient and kind through all that, he felt, and prepared after a decent interval (during which he enjoyed the favours of a pretty little cocotte called Marguerite) to reassume his marital role. But motherhood seemed to have released something animal and sensual in Sarah. It was absurd, and *she* was absurd. If she had been a *cocotte* he might have become quite excited about the way she responded to him – she had even tried to *initiate* things in the bedroom. But a married woman, mother of his daughter – ugh! It was just not done. Only the need for a son and heir had kept him sleeping in the same bed as his wife, and once she was pregnant again he had been highly relieved. Yet when he moved out of her bedroom, ostensibly to allow her to rest and prepare herself for the coming confinement, he knew that she had seen the relief in his eyes and she had looked pained.

René was not a man who intended a large family; he did not much like children, so the only way to conduct himself was to leave his wife alone. Most wives would have been pleased, he thought, at his progressive conduct. Not Sarah. He knew she would have liked more children but she was prepared to deny herself them. From the moment he made it clear that he did not wish to discuss what Sarah very timidly alluded to as 'precautions', she had known that she had lost him.

Everyone thought that his friend Demaine was the *coureur de femmes*, and yet Guy was seemingly quite content to stay faithful to Kate so long as he had an amusing life. No, Guy's trouble was gambling and he knew Kate was careful with her fortune and allotted him only a small allowance to pursue his pleasures. He had a good time of course, old Guy – flirted with young women, but never, so far as he knew, got them into bed. The trouble was, thought René, as he prepared for this boring afternoon – his wife's idea and he had to go along with it – Guy liked his social position and liked to feel envied and courted, whilst he was a more private person and had tastes that were not so easily satisfied.

Marianne and Sarah had finished their decoration of the tall Christmas tree that reached to the height of the gallery door. The male servants had been standing by with step-ladders to receive orders, but it had been Sarah who mounted it to fix the star at the top and then the higher baubles, green and gold and silver. The men had been detailed to pile up the parcels on a table by the tree, each inscribed with a name by Sarah. Marianne was helping Sarah with the various games they intended to have. Surely the village children would know Oranges and Lemons? If not, they would be taught it.

The maids had prepared a long trestle table running down the side of the gallery at the back of which plates and cake-stands and cups and saucers and glasses stood in white and silver ranks. 'I want them to remember this afternoon,' Sarah had said to Marianne. 'Something beautiful,' said Marianne. 'Yes, they probably will.' Beautiful things do not happen by accident, she thought. Someone has to prepare them. There would be other parties for their own children and Kate's, but this was probably the only party the village children would attend. The Revd Ferguson was to come with his flock. Most of the children were the younger sisters and brothers of those children Sarah had once taught in Sunday school and later there would be a sprinkling of Warrender children and others, the scions of some of the very girls they had come out with and who now lived peaceful lives in their rapidly filling nurseries. Would she and Sarah and Kate have more children, she wondered, as she added Musical Chairs to her list. Hopkins could bring in the servants' chairs and Ned go up to the attics to fetch down what Sarah called 'Regency stuff', no longer used and a little rickety.

'Will you be able to get them all to behave?' asked Kate as she came in before luncheon, having had a few polite words with René Boissier, who was standing silently in the conservatory looking rather discouraged.

'Oh, yes, the villagers will be petrified at first. Mr Ferguson will see that they behave after that, when they've stopped being frightened.'

'I thought Ring-a-Ring-o'-Roses for the little ones after Oranges and Lemons,' said Marianne, hovering with a pencil.

Kate looked amused. 'Do you teach Sunday School too, Marianne?'

'No, no – but I've helped a little at the village school. You know Ned was one of the founders of the new National School in the village and we take an interest and even go to their occasional celebrations. They know all sorts of queer games up there, you know. Ned thinks they derive from sheep.'

'From sheep?' echoed Kate, puzzled.

'Yes, funny counting games like Bo Peep, but rather fierce, with lots of running and disappearing. Quite difficult to control.'

Marianne crossed off the last item. 'Well, then, there's tea or lemonade and the biscuits and cake and favours and crackers – they'll take some time. Then Father Christmas – Ned, you promised – you will appear, won't you? There's an old Misrule robe Sarah found up in the attic that would just do.'

'Oh, I've forgotten the mistletoe,' cried Sarah. 'I must go and remind Nanny.'

The holly had already been fixed around the panelled room by the footman and his helper. Lady Gibbs still kept him on although he was no longer young and his former tallness had become bowed. Still, he was willing to do anything asked and this afternoon there were plenty of jobs for the asking.

'Do you think Charades?' asked Marianne of Kate, who had moved to the window.

Kate was remembering Hunt the Slipper, that absurd game they had played here ten years ago when she had talked to Ned in the conservatory. Here he was now, Ned Mortimer, quiet and solid, standing by his pretty wife. She watched them under her lashes and saw how they seemed at ease with each other. Not like Sarah and René, who seemed to have some smouldering problem pursuing them. Perhaps it was her imagination, but she was usually right.

'No, I don't think Charades,' replied Ned, since Kate made no answer. 'Too hard and too much trouble. Needs people who like showing off.'

'Louisa would like it!' said Marianne.

'I don't think Lou must be encouraged – after all the party is mainly for the village, isn't it? They don't want to feel out of place having long words thrown at them. They want to eat and drink and have a present and be with their mates.'

Marianne laughed. 'Always commonsensical. But they have to play at something.'

'You'll find the time will pass quite quickly,' said her husband. 'I believe luncheon is served.'

Marianne put her pencil and notebook away carefully in her

jacket pocket. They had probably prepared enough. Kate swept out before them and was in time to see Louisa hiding behind a velvet-curtained door.

When the grown-ups were safely out of sight, Louisa and Miles came out from behind the door. They stood for a moment irresolutely. Louisa had cold feet. Would it all be spoiled if they saw the tree before the others? But Miles was saying, 'I wonder what games they'll play? I like the English and American games best that Mama brings from London and that Aunt Chauncey brings from Boston, but you can't play that sort of game with fifty people.'

They sat for a moment together in the window-seat in the entrance to the Long Gallery. All was silent, except for a rattle of plates from below. 'What sort of games?' she asked.

'Oh, card games and board games. Kitty likes building with the dominoes, but I play them properly with Mama – and we have a bagatelle board and snakes and ladders, but that makes Kitty cry.'

'I like draughts,' said Louisa. 'But Papa says he will teach me chess.'

'I play that, Uncle Adolphe taught me. I can beat my Papa at that,' said Miles.

'And Halma and Patience and Snap, and Beggar my Neighbour – '

'And Noughts and Crosses and Happy Families and Tiddly Winks – '

'And Spillikins and Conkers – that's a boy's game, but I like it – '

'And we had a new jigsaw puzzle from Boston too – '

'Yes, Hugh has one – '

'And Mother plays Cat's Cradle and Cribbage – '

'Cribbage? What's that? I don't know that.' This boy seemed to know everything. Louisa was cross she could not yet play chess either. 'Will you teach it me? And to play chess?'

'All right.' They were both somewhat breathless after their rapidly fired catalogues.

'I wish we could just go in the nursery and play Lotto or have a spelling bee,' said Louisa loftily. 'None of the party games will be interesting.'

'We can afterwards. But you'll enjoy the party.'

'I don't know. I'm not good at Musical Chairs and things like that.'

'That doesn't matter – we'll have to let the village children win, won't we?'

'Shall we? – yes, I suppose we ought to.' She had not thought of that. This boy was very grown up. 'Did your Mama tell you we mustn't try to win?'

'No, no – Mama always likes to win at cards. I just thought, you know, we've probably had more practice than the others. Shall we look at the tree, Louisa?'

'All right.' They both looked round and then crept up to the door and opened it cautiously. It was not very bright in the room and the light outside had begun to fade even then, but the snow-light came in through the long windows and they stood for a moment irresolutely and then saw the tree at the far end, all tangled with skeins of silver and with great green lanterns and gold stars, the largest star at the top. 'Oh,' breathed Louisa. 'Isn't it lovely?'

They were both whispering.

'Look at the presents on the table.' There was a heap of little green parcels in the gloom.

'Don't let's go any further,' muttered Louisa. 'Let's wait to see the tree properly.'

But Miles stood now at the long window looking over the snow with a sky of sullen pewter and sepia above the whiteness. He said nothing for a moment. Something about this English landscape made him at once sad and excited. Yet it was all so ordinary. 'The snowman – doesn't he look good!' pointed Louisa. '*We* did that. I hope he'll stay till tomorrow.' With common consent they moved back through the door to the corridor.

'We must ask for a pipe and scarf – we forgot,' said Miles in a normal voice.

'Don't tell anyone we looked,' urged Louisa.

'All right. Race you to the grub.' Where had he heard such slang, she thought. He was really just like a completely English boy.

Aunt and Uncle Leverton were to arrive later that afternoon, so Lady Gibbs was much preoccupied and was relieved to leave the children's party to her daughter and her friends, with the help, of course, of all the servants, who did not all approve of having the village attend. They themselves did not consider that they belonged to the village but to the Hall – there was a difference after all. But they did as they were bid with a certain amount of grumbling. Meanwhile the children had had a small luncheon and rested and they were now being dressed in a warm nursery, Sarah's old one. Her brothers were also to arrive that evening from their various pursuits and she was determined to prove to Roger and Ralph that she had achieved a successful afternoon for all. Not that they were expected to join in. They were more on their dignity than their sister, always had been.

Miles and Hugh and Valéry were given a lick and a polish in René's dressing-room where stood the large ornamental wash-stand that had been in the house as long as anyone remembered. It was cold outside and the skies promised more snow. Miles was philosophical about the party. Without conceit he regarded the principal problem as that of not winning the games. He was a tall, solidly built boy who looked reserved, but upon further acquaintance was found not to be. Hugh Mortimer had already marked him out as an ally against his sister's bossiness. They stood submitting to the *bonne*'s administrations. Miles was used to velvet knickerbockers for Kate dressed her children fashionably. In the nursery little Kitty was last to endure the hair-brushing and the tight combs and Alice band. Kate had bought her a red dress in Paris and it suited her dark hair and dark skin. The ribbons were also scarlet and she wore a silver bracelet. Louisa rather envied the dress, although she too was to wear her best party dress, brought from home wrapped in tissue by Polly who had been busy sorting out chemises and stays and white underpetticoats and flannel petticoats and the freshly washed and ironed drawers with their open scalloped edges. Polly thought Louisa looked a picture in the white silk dress with its blue ribbons matching the ones in her hair. She was being allowed to wear her shawl too, a silk shawl of noonday blue. Black pumps completed her outfit, whilst Kitty had a scarlet cloak and Emma a little cape edged with fur. Olivia, the tallest of the girls, was rather nervous. She knew that her mother was counting on her to play hostess to the children and to cut the Christmas cake and she felt rather uncomfortable in her best dress, made in Paris, but somehow never quite the right fit for her long English waist. She was also wearing a new pair of boots, having outgrown her dancing pumps, and they were uncomfortable. But she decided to make the best of it. When she was grown up, she decided, she would wear long gloves, glacé boots and pale ribbons. At present they were always sliding off her ringlets. Papa wanted her to look nice and be a credit to him. He had even bought her a set of silk lingerie, a somewhat unusual thing for a Papa to do, but very beautiful. She was wearing a coral necklace which Grandmother had given her on her last birthday in the summer.

'Let's look at you all in the glass, then,' said Nanny and put them in a row against the floor-length glass on the top landing.

'It's got marks on,' said Kitty. 'It makes my face look dirty.'

'Never mind that, that's old mirrors,' said Nanny wisely, giving them all a final critical glance and adjusting ribbons and tweaking hair.

'We shall get very untidy,' said Louisa, 'playing all those games Mama has arranged. Hadn't we better have our Cabinet pictures taken now before.'

'I don't think he's come yet from Oxford,' said Nanny doubtfully. 'We can always tidy you up again, Miss Louisa.'

They were all to wait in the small sitting-room until the guests were brought round to the Hall by Mr Ferguson and the others arrived with their maids or Mamas. The boys were already rushing down from their confinement in the dressing-room and Polly was put to sit with them all. Parties were quite terrifying, decided Louisa. Perhaps roller-skating would be better. Still, Miles was there, and that was fun.

The chatter of children outside the great porch (for at Sarah's insistence they were being allowed in at the front) reminded Marianne of something she could not at first identify. Then suddenly she remembered. It was her wedding day when her sisters and brothers came underneath her window waiting to walk over to the church. How long ago it seemed. Of course, ten years *was* a long time. But it seemed strange that now there were other children – a new generation. How swiftly these generations trod upon each other's toes. They were indeed 'hungry' as the Bible said. It was only occasionally that you could stand back from your life and see it through the other end of the telescope. Usually one was involved in a more microscopic way of looking, she thought. Worried about a child's fever or about money or servants or whether one was suitably dressed. Why could one not stand back more often and savour it all? Marianne shook her head slightly at her reflection in the mirror and went to see if her own children were ready and looking tidy and presentable. Polly had no time for concentration on detail. Still, Polly was a find – cheerful, cynical and hard-working, a real Northerner. What would she do without her? Three children were a handful.

The children were all waiting with Nanny Hopkins and Maud, with Polly hanging back a little at the top of the front staircase which led down into the great hall of the house. They stood like soldiers, Marianne thought, as Kate and then Sarah came from their rooms, the one shivering slightly and pulling her elegant shawl round her elegant shoulders and the other flushed and carrying a basket of last-minute presents. Everything seemed to be ready and now the great bell clanged.

'The curate is here – they're waiting at the door,' said Marianne and took Emma's hand and followed Sarah down one storey. The other children followed, not soldiers now, but dragging their feet slightly as if on their way to execution.

'Cheer up,' said Kate, casting an amused glance at her own daughter.

Only Miles looked in charge of himself. Louisa looked at Miles. He would know what to do, so she only had to follow him and everything would be all right. They went to the gallery door which was slightly ajar. The room looked now more like a ballroom, the tree glittering at the far end.

'They won't light the candles yet,' whispered Miles.

Just then the curate and his long crocodile of children were heard coming from the downstairs cloakroom. Plod, plod, plod. Louisa giggled. Sarah was already with the Revd Ferguson. Lady Gibbs was to do the honours. She opened the doors of the Long Gallery and stood there for a moment looking at the tree. Then she moved aside as the village children passed into the room and watched their faces when they too saw the tree. Some looked cowed, others nonchalant and several little girls were pale with apprehension.

'However are we to get started?' asked Marianne.

'Play Oranges and Lemons,' said Miles. 'They'll like a bit of noise.'

There was an old grand piano on the same side of the Long Gallery as the tree. Marianne had promised to thump out some easy tunes. Nanny Hopkins and the French maid and Maud and Polly were drawn up by the door to the cloakroom, looking ready to chop off a few heads. They curtsied to the curate when he led in his party. Sarah moved to the middle of the room. There seemed so many children and all so quiet. A few held hands. Others looked upwards, but at the sight of the tree there was, thankfully, a general 'OOOHH!' Marianne consulted with the curate and he arranged all his charges on the chairs that lined the walls. Louisa and the others sat at one end till Miles moved away and she followed and sat down near the newcomers. The younger children did the same and soon Emma was talking to a small girl in ringlets, the daughter of the Post Office lady. The boys preferred to keep mum and the rest of the little girls, dressed in white pinafores and long, straggly cotton dresses of varying hues, stared at Louisa and Kitty and Emma. But now Mr Ferguson was explaining the game and Marianne moved to the piano.

'Some games, children,' he announced. 'Before we are to be given tea by our kind patrons, Mrs Boissier and Lady Gibbs.' He bowed to them both and started to clap. The children clapped dutifully. Marianne knew that excessive shyness often led to excessive high spirits when the initial emotions had worn away, so began to play quietly while Sarah, used to Sunday School outings in her youth and whenever she was home, divided the children into teams, one headed by Miles and the other by Louisa. The music began and Ferguson's hearty baritone began to sing. A few children joined in, but one boy pulled at Marianne's sleeve as he passed the piano. 'Miss, will there be presents if we win?' 'Wait and see,' said Marianne. He rejoined his mates with renewed vigour. It would be a good idea to give prizes for the first game, she thought. Then there might be more enthusiasm for the others!

But soon the children were away and singing and as the great room darkened, the maids came in with tea and sandwiches and stood waiting as the game progressed. Louisa was wondering what sort of games the village children played. She tried to talk to one or two of them, but only the Post Office lady's daughter was forthcoming, and the son of the village schoolmaster, who won the first game and was set to win the next until it was announced that you could only win once. He looked disdainful and angry at this and Louisa did not blame him. Soon there was much more noise and the men, attracted by it, came up. Ned and René and Guy stood at the door watching, René with a bored expression, Guy laughing and occasionally pointing out to Kate (who was seated by the fireplace behind a screen toasting herself) the antics of the assembly.

'Sarah is so good,' said Kate.

'It seems to be going well.'

Indeed, Musical Chairs was so popular it was requested twice and Marianne played the music of Sir Roger de Coverly again and again.

'Now we have a surprise for our kind hosts,' announced the curate after consultation with Sarah. He moved the piano after putting his flock in the charge of the oldest girl, a large, stout child of about twelve, who immediately put the stragglers in order.

'What is it?' asked Marianne, coming up to the fire at the other end, glad to be released from her pianist duties.

'Sellenger's Round,' replied Sarah.

The Hall children stood and watched as the village performed its oft-practised country dance. They clapped it very politely.

Nanny came up. 'Tea after the next game, Madam?' she asked Lady Gibbs, who asked Sarah.

'Please dance it once more and then we shall have some tea,' she said in the silence after the dance was over, and the curate stood and bowed. So Sellenger's Round was shuffled and stumped and occasionally danced gracefully through once more. Guy and Ned advanced towards the tree. Now they all saw that gas candles were fixed on it and Guy, who was good at this sort of thing, illuminated the tree as the dance finished.

'An orderly line, children,' called the curate, and the children, pushing and shoving now, were arranged in a long row leading to the white festooned tablecloths where the large tea urn and the plates of small sandwiches and cakes awaited them.

'Can we eat too?' asked Emma. 'I'm hungry.'

'Course you can,' said Valéry.

'At the end of the line,' said Marianne.

It had not been foreseen that some of the children would find difficulty in carrying glasses of lemonade or cups of tea to their lips, and there were a few accidents cleaned up by Polly, but on the whole things went well. Louisa determined to sit next to Miles at the long table and found on her other side Florence, the post lady's daughter, who seemed a little more *au fait* with table manners than some of the others.

'Not all the boys wanted to come,' the curate was confiding to

Marianne. Marianne was not surprised. It was a little like the feeding of the five thousand, she thought, and she was about to say so when Ned came up.

'Tea for the organisers behind the screen,' he said.

Marianne found Kate and Sarah standing there with teacups and sandwiches.

'Is it going well?' Sarah was worried.

'Fine,' said Kate. 'Do you mind if I escape for a moment? You could perhaps ask Marie if Kitty is tired?'

Sarah then strode out to the far end of the room by the tree while the assembly ate and drank. She began to check the presents which the children would take home. It was quite dark now outside and she looked back at the room wishing René would come up and congratulate her. But he seemed to have disappeared. Lady Gibbs came up.

'It was a good idea, dear – almost as difficult as a coming-out ball!'

'Better behaved though,' replied Sarah.

'I did not know the English were so democratic,' said a voice at Marianne's side. Turning, she saw Guy at her elbow.

'How well your son behaves,' she replied as Miles was seen to attend to the needs of a small village child who was wailing for 'anuvver pop please'.

'Yes, but of course your Louisa is so grown up, too,' he replied politely.

Marianne's heart leapt in her chest, but she coughed to cover it and said, 'I must go and see if Sarah needs my playing for the next game.' He followed her with his eyes in the gloom and smiled at Louisa, who was sitting at the head of the table with some small ringleted child on one side and Miles on the other. René returned and came up to his friend.

'Kate has a slight headache,' he said. 'She sends her excuses.'

Guy laughed. He well knew that Kate felt she had done quite enough in the cause of English tea-parties. *He* had done more. He did not suppose that many Englishmen, or Frenchmen for that matter, actually attended children's parties unless as an act of charity.

'Sarah has not changed,' he remarked. 'She'd give a party for horses if she thought it would please them.'

René smiled and stroked his short beard. He was more reserved than he used to be, Guy thought. Ned, though, was enjoying himself, looking jovial and handing round crackers now with glee.

'Hey, mister,' said a boy. 'I ain't got a piece of cake.'

Ned remedied this and then sat down next to Louisa who made a place for him at the table.

'Isn't it nice, Papa?' she said. 'Do you think they are enjoying it?' she whispered when he pulled a cracker with her.

All the little girls were exclaiming at the favours in the crackers and a few boys were trying to bang them in the ears of the girls, so Mr Ferguson, after consultation, thought it best to announce that grace would be said and then there would be more games. The children sat and shuffled through the short 'For what we have received' and Sarah took charge again.

Marianne stood now looking over the park at the darkness. She hoped Guy would not approach her again. It was true, though, Miles was an awfully nice boy. René, too, was looking out at the park and then making some remark to her which at first she did not hear.

'Louisa is very pretty,' he said again. 'She'll be a beauty.'

She smiled at him. René was wondering of whom Louisa, her face now flushed and eyes sparkling, reminded him, but he could not think.

'I expect a lot of the children don't get square meals,' she said. 'With the agricultural depression, you know. Sarah says it's getting worse.'

'They don't look too bad to me,' replied René.

'I don't expect the worst cases go to church,' said Marianne, 'so they might not be here.'

The children had been marshalled for Here We Go Round the Mulberry Bush and a little posse of girls was asking for Ring-a-Ring-o'-Roses. Thankfully now she went towards the piano again.

The games were all played and the level of noise was rising once more, when Sarah finally clapped her hands and announced that Father Christmas was about to appear. A hush fell and a tall, red-gowned figure appeared at the far door with a large sack. He sat down on a chair and delved into the sack, helped by Sarah, who read out each name separately. Some of the children giggled, others jumped up and down and Mr Ferguson stopped a few imminent fights. In another sudden silence Ned in the red robe began with Teresa White, Edith Herbert, Bessie Thomas and on and on till the girls were done. Then it was Benjamin Woodford, Herbert Smith, Albert Hughes and, finally, after each child had received its gift, the seven children from the Hall party. Hugh looked inquisitively at the red-faced and red-cloaked stranger and thought the whiskers rather like Papa's. Now all was chaos and shouts till Mr Ferguson asked for 'Hip, hip, hurrah' and the children obliged shrilly.

'And, children, I hope you will take away for ever the memory of this fine party,' he went on. 'Look once more at this tree, symbol of Christmas and remember the message of our Saviour; that it is better to give than to receive.'

Marianne was wondering where in the Bible was this admonition as the final Hurrahs were echoing and the children, about to leave, were standing in twos and threes staring at the tree. They probably *would* remember it. It was beautiful. Now the only lights in the room were from the tree and there were shadows in the corners. The Hall children and guests went down with the village children to wave them goodbye across the park and she stayed there a moment listening to the winter sounds as a full moon began its steady rise in the tree-enlaced sky.

Sarah was exhausted and came back with her children to sit by the fire with the screen now removed. Kate returned with a velvet neck warmer and sat for a moment too as the Hall children, one by one, returned. Then the servants were allowed to take their own tea at the table. The men had disappeared somewhere, probably to the billiards room.

'Wasn't Ned splendid as Santa Claus? said Sarah yawning.

'He likes dressing up,' laughed Marianne.

'He is a good father, isn't he?' said Kate. 'Not just a Father Christmas.'

Marianne was a little surprised. Kate rarely spoke either to or about Ned. 'Yes, but I am not as good a Christian as Sarah,' she replied. 'Sarah – all your work! But they did enjoy it, they'll be showing their mothers their presents now. It was magical.' Sarah looked pleased.

Now the Hall children were rushing up and down the gallery, their mixed emotions at last released once the agony of having to behave well was passed.

'Wish we could skate here!' cried Valéry, seizing little Kitty Demaine and rushing her up and down, their feet skittering on the wooden floor.

'We'll have to put some French powder on it – ask Thomson', said Lady Gibbs.

'I'm a Rollerama, I'm a Rinkomania,' cried Hugh.

'A Rinkomaniac,' corrected Miles, chasing him down the galley.

Louisa crept near the fire. She was not cold but she loved the glow. Did the others feel quite as she did? The tree's lights, all that lovely jadey green and the silver reflections and the sight of wrapped-up presents releasing a complicated mixture of rapture and something else. She struggled to find a word for it. It was beautiful, but it was also somehow sad. No – that was silly – yet there was something sad about an event, a tree, colours and excitement, that lasted such a short time and then disappeared for what seemed years and years, though it was only twelve months.

Miles was off now, playing some boyish game or other and she saw her little sister Emma creep up to their mother and nestle against her. Father had been Father Christmas of course – he hadn't fooled *her* one bit. Anyone could have seen it was him with his blue eyes and his smile. Oh, she did love Papa – and Mama too – and Aunt Sarah who was such a kind lady and Aunt Kate who was mysterious and Uncle Guy who was such good fun and Uncle René who had looked at her in a special way and said, 'You're a pretty girl, Louisa Mortimer,' which had made her blush.

'Wasn't that little girl's dress pretty? – the one with the pink frills?' she said dreamily to her mother.

Olivia answered her: 'It was her sister's,' she said. 'I saw her wearing it last year. It was a little too big for her.'

'Oh, was it?' said Louisa in surprise. Fancy Olivia noticing.

'Pink organdie is not very practical,' said Kate, overhearing and toasting her feet. Soon Nanny and the other servants would be bearing away the children and they would all be expected to go, she supposed, to the midnight service at the Revd Ferguson's church across the park. She did not want to go, but her Catholicism was no longer an excuse since she had not attended mass since her marriage.

Olivia and Louisa and Miles were allowed to stay up for a piece of Christmas cake and a sip of wine that evening while the younger children were bathed in the nursery. The grown-ups then prepared for their carriage drive to church.

'I wish I could go,' said Louisa to Miles. 'I like singing – specially the Christmas hymns.'

'Mama is exhausted,' said Olivia. 'At least I heard Grandmama telling her she was. But Mama is very strong.'

When Sarah did look in for a moment at Olivia and Louisa, who lay in bed in the nursery firelight with the stockings neatly folded awaiting Father Christmas once more, Louisa said, 'Thank you, Aunt Sarah, it was just splendid.'

'Why, thank you, Louisa darling,' said Sarah. 'But there's still tomorrow. It isn't all over yet.'

'Tomorrow and tomorrow and tomorrow,' mumbled Louisa to herself when the grown-ups and Polly and the others had all gone downstairs.

Days whizzed by like shooting stars. If only the holiday would go on for ever and they could stay and she could play with Miles and beat him at a spelling bee. She would allow him to beat her at Snakes and Ladders.

There was a 'Rollerama' down the long servants' corridor on Boxing Day. By then the children were all over-excited and their parents were already working out how long they could stay. Sarah was to stay a while in England before rejoining René in Paris. 'It does you good being here,' he said truthfully. For Sarah had never really acclimatised herself to Paris and it was a lesser expense for the children to stay with their grandmother while the Parisian apartment was empty of all but himself. She was longing for a few long rides – she would go hunting, of course, but preferred to ride alone to outflung villages calling on old neighbours, which bored him.

Her 'religion' always made him feel guilty too, for in England religion meant that you also possessed a social conscience. She had consented to the children being baptised as Catholics as he himself had been (though now it was only a nominal appellation as he fancied himself a Rationalist) and so long as religion did not interfere with his pleasures, Sarah could believe what she wanted. But the apparent intensity of her faith made him uneasy though she never spoke of it to him. Sometimes he was sure she had been praying for him!

'Your English women are so virtuous,' he had once said to his wife with a slight sneer. Sarah had looked at him sharply but said nothing. He let her go out with her soup or whatever it was she took to the ailing villagers without being followed by one of his cynical remarks. She had had her children's party too and he had tried to be pleasant. He had nothing to reproach himself with but, *mon dieu*, he was bored! There was no one to flirt with in this great cold place and he had had to talk to his mother-in-law for what seemed like hours. Still he had done his duty and everyone else seemed happy. Sarah looked happier too in England.

He did not dwell on that faint fleeting resemblance to someone he could not name that he had seen on the Mortimer girl's face, though he had automatically registered its prettiness. He had other things to think about, chiefly his new mistress who was being difficult and wanting him back in Paris.

Kate and Guy were to return to France in the spring where an even grander apartment was at that very moment being prepared for them. Kate was content. She had decided to settle in France now that the 'unpleasantness' seemed to have passed. Guy was sometimes a little too casual in Italy; they must now make their mark in a more sophisticated milieu.

Ned went North the day after Boxing Day and Marianne was to follow, along with Polly and the children. Before that they gathered at dinner and on occasional walks, but all felt a little unsettled now the parties and Christmas celebrations were over. 'The sadness of Epiphany', Marianne thought. January and February were harsh up in the North and she must find a new governess for Louisa and perhaps a school eventually. Hugh was to go off to his father's old prep school and only Emma would be a nursery child. The future beckoned her as it did them all, but she hoped nothing would fundamentally change in her life, for she was quite content now. Even rereading Meredith's *Married Love* she had wondered what all the fuss was about. Was it not better now she was nearly thirty to settle down to some peaceful thinking and calm outdoor pursuits? She did not really envy Kate her glittering life nor Sarah's opportunity of 'swimming' fashionably if she wanted – Sarah seemed a little unsettled after Christmas and confided to Marianne that there were problems with the land that neither Roger nor Ralph had foreseen. Marianne, who had grown up amidst grumblings and fears of collapsing businesses and lack of ready cash, was worried for her friend. Lake Hall Park could surely not change. It must stay for ever as it was.

At one of their last dinners *en famille*, when the men, except for

Guy, had gone, Sarah's guests discussed the new Divorce Laws and the effect they had had on the country. A certain Lady Ebson had had a divorce petition filed against her for adultery and, of course, it had been the talk of the neighbourhood. Marianne avoided talk of adultery. It had never entered her head to be unfaithful to Ned and she was sure it had never entered his either, but Kate seemed interested in the fact that simple adultery by the husband had to be joined to some other cause for a husband to be cast off by a wife. No one mentioned what crimes or causes were, in fact, needed for a woman to be free of an errant husband, but Marianne again felt uneasy.

Kate said, 'My Papa deserted my Mama, you know, long before she died, but she could not divorce him.'

There was a silence. Kate had never mentioned her parents before as far as Marianne remembered.

'Of course,' Kate went on, 'they would say now there was a reasonable cause since Mama was not an easy person to get along with.'

Was she challenging Guy to something? She looked at him boldly as she said this, but all he did was smile and say, 'My dear, many men commit adultery, but that does not make them into monsters.'

Marianne wondered then whether he, too, was unfaithful to Kate. She did not think he would be. It was surely more likely that Kate would be unfaithful than Guy – but one never knew. Men were unpredictable. Even in the provinces there were goings on that could have led to divorces in more sophisticated sections of society.

A neighbour who had been invited to make up the missing numbers of men, said, 'Down here we care for the sanctity of marriage.'

Marianne pondered the sanctity of marriage all through the pudding course. Was she not a fraud calling herself a Feminist when she relied upon Ned's work to feed her and her children?

The conversation turned to food and she breathed a sigh of relief. Guy was saying rather rudely that Americans could not cook.

'I don't try,' said Kate laughing. 'We have a good chef in Paris, one not so good in Italy and my husband is full of complaints.'

'René says only the French can cook,' said Sarah.

'Well, if the ladies give up on their cooking what are we poor males to do?' asked another neighbour.

But, thought Marianne, the ladies here did not cook: they only ordered other people to do so.

Marianne turned to a younger man, a friend of Ralph Gibbs who was sitting silently at the end of the table consuming claret and not listening to the conversation. 'Have you read Frances Cobbe on Dreams?' she asked. 'It is a most interesting article.' The man looked nonplussed. She tried again with 'We have ladies who lecture on such subjects up in Manchester.'

Lady Gibbs caught the drift of her words. 'My dear, Lady Russell read a paper at the Mechanical Institute in Stroud,' she said. 'About suffrage, I believe. The county was up in arms.'

'Mama, I didn't know you followed such things,' exclaimed Sarah in surprise.

'Well, dear, one hears about the Shrieking Sisterhood – such a vulgar title, I feel. Are they all American, Kate? One hears such things about the States, what with negroes and Mrs Beecher Stowe – although of course that was some time ago. But after all, we are all children of God,' she concluded, somewhat vaguely.

'Against nature,' said the first neighbour on Marianne's left. 'Ought to be put a stop to.'

Marianne decided to change the subject once more. What did they usually talk about here? She missed Ned and the rationality of his conversation and even missed René who would have quipped something amusing and led the conversation to something lighter. She was no good at dinner party conversation, never had been. The ladies finally rose and assembled in the small Pink sitting-room where Sarah's neighbour, the wife of the anti-suffrage gentlemen, was discussing the purchase of a vivarium for her children. 'We need exercise,' someone said. 'Have you played the new Badminton?'

'My dear – not after dinner.'

'No, well, you must all come to the Grange, we've a good battledore and shuttlecock court. It's easily adapted to Badminton.'

The talk then passed to magic lanterns and the Camera Obscura game. 'Miles likes that,' offered Kate, coming to life. 'He's so interested in technical things – says he wants to be a photographer, but there's no money in it, is there?'

Marianne thought the conversation dull. They might just as well emulate the children and play Bagatelle or Floral Lotto. But then

Sarah indicated a chess board and Marianne found she was challenged by the anti-suffrage gentleman who was called Cedric, and agreed to play. She was not very good at the game, though she enjoyed it, but surprisingly this evening she began with an unusual queen's pawn opening which flummoxed the Cedric person and she was well on the way to winning when the neighbours with the large carriage decided they were to leave and Cedric (thankfully, she thought) abandoned his game with her.

'Happy Families,' said Kate irrelevantly when they had gone. 'I expect the one who finds divorce so distasteful is cheating on his wife and the one who is against women's emancipation is henpecked.' They all laughed, but Marianne thought, I wonder what Kate really thinks or feels. I don't know her any better than when we were nineteen, though now she is more friendly to me.

The next day the women went for a walk in their winter boots across the fields where the snow had now melted. It was pleasant country, but Marianne thought it too tame and Kate did not enjoy walking anywhere but on city pavements. Once the men had gone or were absent and the talk was not clogged up by neighbours, the women found it easier to spend the last two days of their holiday together, knowing the children were well looked after in the nursery or still roller-skating up and down the corridors with shrieks and bumps and some tears. I really must devote more time to Ned, thought Marianne. I miss him when I am away from home, but at home I'm never lonely, even if he's busy. I wonder why.

Kate thought, I am beginning to be a little bored. I ought not to have said that about my Mama the other evening, but I wanted to see their faces. I wish I did not find children such a drag.

Sarah was thinking, How did I let René down? I am sure he is with Bella again, or someone worse. Is there something wrong with me? But she could not think of one reason why he acted as he did. It never occurred to her that there was a distinction in the minds of some men between 'pleasure' and 'love' and that the fault was not hers. She knew she was not beautiful but after she had recovered from the two births she was quite passable. She knew this, since one or two of her husband's friends had allowed their hands to linger in hers and had looked at her in a way that meant only one thing. But she found that truly immoral. She was René's wife and the idea of an extramarital attachment was disgusting. But if others found her even a little attractive, why did her husband not find her so? Gradually she had come to believe there was something wrong with her feelings, was ashamed of her erotic dreams and could not help continually praying that René might find her desirable as she thought he must once have done. But there was no question of any deep emotional relationship between them. She had missed her chance of true and mutual love so she would strive to love him in the only way left open to her. She would be loyal and kind and efficient and reserve her deepest feelings for God, her children, and her horse, a black mare she kept in a stable near the Bois. She would not allow herself to grow bitter, would even pray that she might gradually lose her carnal desire for her husband. Perhaps this was what happened to all married women? Once or twice she found herself thinking of Ned Mortimer and for a fleeting moment would wonder what it would have been like to marry him. But he was so ideally happy, you could see, with Marianne.

'You must come North,' said Marianne to both of her friends. 'The daffodils are a magnificent sight in April – all along the lakeside. It is wonderful air up there, all the children would love it. *Do* come – or send them.'

Why does René need other women, Sarah went on and on thinking, even as she answered Marianne with 'That would be splendid, dear.'

My money and his charm, Kate thought. Together we shall storm Paris as we stormed Florence and Rome.

I can tell Mama, thought Sarah. But she must know all is not well, or why should I spend so much time away from René. I wonder if Papa . . . ?

'And in the summer there are rambler roses, hundreds of them, in our lakeside garden,' Marianne went on. 'I have tried to paint the scene. And in winter we have the great log fires from the pine trees up on the fells.'

Guy is weak, thought Kate. But not weak enough to make any silly mistakes. She smiled to herself.

They came home to the Hall, each with their separate thoughts. Two days later the Demaines and the Mortimers and their servants left in a flurry and scurry of children and baggage and trunks and promises. Sarah waved them goodbye and went in to her mother.

It was to be some time before they would all meet again. Louisa, whose party had left first, had waved and waved to Miles until his figure, in the carriage courtyard, had disappeared from sight.

Act Two: Reaping

1886–1890

14

Louisa Mortimer was almost nineteen years old and not at all desirous of coming out in society. Whoever bothered about that in Westmorland, except the daughters of the nobility and a few girls who aspired to become wives of the same? Louisa was scornful. She agreed to come out in her own way by accepting an invitation to stay in Paris with her Aunt Sarah Boissier, whom she had not seen for some time. Louisa was desirous of attending one of the new ladies' colleges in Oxford, for she had had a better education than her mother at a new Girls' High School founded in Manchester when she was about twelve. Her sister, Emma, had been supposed to follow in her footsteps, but had disliked being away from home.

'Paris will be useful, Papa,' said Louisa. 'I have still to pass my Latin Responsions, you know, if I am to study French, but it will be good to practise the language with Valéry and Olivia, I'm sure.'

Ned could never refuse this lively, passionate daughter of his anything, and agreed with her that the season was a waste of time. 'Though your mother and I met in it,' he added.

Louisa was very fond of her father, and found him easier to get on with than her Mama. But Mama had made no objection to her visiting Paris so long as she took the opportunity to study and not just gad about. 'As if I would,' objected Louisa. She had always been a rather rebellious child, thought Marianne, and the spectre of this rebellion turning to an unsuitable attachment (even if she tried not to think about the *most* unsuitable form that attachment could take) had always worried her. Louisa was as headstrong and impetuous as she herself had been, she supposed, but in Louisa's case, it was allied with less desire to please her Mama and a greater sense of her own righteousness.

Sarah had reported that Kate and Guy's son Miles was at present in Boston. Kate was in Paris, but was talking of visiting the States soon. Sarah suspected that Guy had found the gaming tables of Monte Carlo too much of a magnet in recent years. Kitty, at only fourteen, was reported a beauty and likely to follow her Mama as Queen of American society abroad. It was a little of a pity, thought Marianne, that Louisa refused to have a proper season, for even if she did not want to meet eligible young men she might show Marianne's old acquaintances that *her* daughter was just as beautiful and probably more intelligent than the Kittys of this world. The same could not be said of Emma, who was much more placid, but who would probably be a success one day, for she was rather obstinate, although shy. She was not so critical, or at least not so openly critical as Louisa, who had said as one of the reasons for her non-participation in the social round: 'Hermione Walker-Brigg's sister said it was horrible having to kiss the Queen's hand. You have to just brush it with your lips, you know, and she said it was like a piece of marshmallow and it was agony trying not to touch it with your nose.'

'I don't expect Her Majesty is delighted at having strange girls touching her in their hundreds,' said Marianne, laughing even so.

'No, they said she was really po-faced and gloomy-looking.'

The money would be better spent giving Louisa a little 'French polish' after all. Ned's business was up and down and he had kept his own tenants on the small farm next to his parents' old house at Langthwaite in case one day the cotton industry collapsed altogether with foreign competition.

Marianne was happy at Langthwaite and well acquainted with all the moods of the Lake District, both sullen and sunny. But more and more manufacturers seemed now to choose the place as a playground or a retirement home, and villas and new settlements had sprung up on both sides of the lake. It was still wild and sometimes dangerous once you left the well-trodden paths, though you might now meet organised parties of rock-climbers. The trains ran so frequently that sometimes, Ned grumbled, the place was turning into a Blackpool.

He and his wife often walked together on the lower fells for they were good companions. Their son Hugh was almost seventeen and doing well at Rugby School, which they had chosen instead of Harrow. He now had a small sailing dinghy on the lake. Emma, however, was being educated at home. She had none of Louisa's high spirits but was happy fishing and swimming with her brother, caring for her animals and sometimes bird-watching with her father and the son of Lord Ireson's gamekeeper and the gamekeeper himself, whom Ned admired for his special knowledge of fauna and flora.

Everyone knew that Louisa was her father's favourite and he hers. Ned saw her mother in her and escaped when the storm clouds drew too near. He was a coward when it came to women's sulks or tempers. Not that his wife often gave way to them. As far as Marianne was concerned she was more likely to be prey to sudden or prolonged anxieties than lose her temper with her calm husband. That Ned loved Louisa so, both gladdened and disquieted her. As the years drew on she tried to think less and less about her past. Indeed sometimes she wondered if she had dreamed it. She was determined that Louisa would not make a mess of her youth and marriage. Not that she herself was not happily married – but she knew she did not deserve it. Louisa deserved the best, and must have it, and it would be a good idea for her to do some further study before attempting the next year to enter one of the ladies' halls in Oxford. She was both musical and literary, but also attractive enough to gain the approval of the male sex, which might or might not be a snare.

Sometimes Marianne would wonder where Kate and Guy were, what they were doing, while she was helping with the orchard apples or wandering down to the village in the dog-cart. By all accounts – chiefly Sarah's – Miles Demaine had grown into a tall, handsome lad, old for his years and thoughtful. She had never set eyes upon any of the Demaines since the children's party at Lake Hall Park nine years before.

Marianne was worried about Sarah. True, Sarah never complained, but she had seen her, on and off, in London when she and Ned had gone for a few days' holiday, and in the North, when Sarah had come with Olivia, leaving Valéry with his father in France to attend a lycée in Paris. Olivia had grown very tall and was rather a cross girl, which Sarah certainly did not deserve. Sarah had been worried about Lake Hall Park. More land had been sold and Roger had decided to go to Australia, leaving Ralph, the less able of the two, to manage the estate. Lady Gibbs still struggled on as more and more of the old plate was sold off at Christies. Sarah had enjoyed the countryside around Langthwaite and had hunted with the hounds on her visit and made a good impression on the local huntsmen and on the farmers with whom she talked unaffectedly of crops and sheep.

Now that Louisa was finally off to Paris to stay with her dear friend, Marianne wished she could accompany her, but Ned had had a bad bout of pneumonia and needed nursing, so she put Louisa into the charge of Alexander Carmichael for the journey. He often travelled abroad with his new patents and was not adverse to accompanying the daughter of his old friend Ned. He took his sister with him too, for she liked to see the sights of Europe while he did his own business in Germany and Scandinavia and, occasionally, Alsace.

'Be a good girl,' said Ned when Louisa came to say goodbye to him. He was lying propped up against cushions. The illness had drained him and Louisa suddenly felt such compunction that she said – a trifle disingenuously – 'Darling Papa! I won't go if you don't want me to. You do feel better, don't you?'

'Off with you, Lulu – I feel much better and it's time you went abroad and learnt how to act in a ladylike manner,' he teased her.

She kissed him goodbye, but felt relieved when she was finally in the train at Kendal with the Carmichaels. Her Mama would have more time to spoil Papa if she were gone. She was young and happy and full of energy and when she returned Papa would be much better.

'Aye – he works too hard,' said Mr Carmichael. 'Your Mama will make sure he rests now.'

'Don't fret,' added Miss Carmichael. They both liked Louisa, who was talkative, even though she sometimes tired them.

Once Louisa had gone Marianne did indeed devote herself to her husband, with the help of Emma who was practical. Marianne had been quite frightened that Ned would not recover. She had never seen him so low. She hoped that the spring would do him good. Already the trees had that faint cloudy green on their branches and the sound of running water from the brook at the bottom of the garden could be heard from the windows of his sick room. Louisa would manage. She must learn that she was not the centre of the universe. Paris would show her other fashions, other ways of living, different from her materially comfortable but wholesome and unshowy North-country home, and she would return and make up her mind what she wanted to do with her life. Paris would certainly be an improvement on Madame Duplessis and it was comforting to think that Louisa would be with Sarah who was of all people the one whom Marianne trusted most.

Louisa was sitting in the inner courtyard of the new Boissier apartment near the Luxembourg Gardens with the spring sun gently caressing her face. On her knee was her French grammar, but she had not looked at it more than twice since coming down after luncheon. French meals were so enormous. Delicious, but they made you sleepy. Especially luncheon, which was much more ample than she was used to. Paris was beautiful, but she so missed trees and flowers that she prevailed upon Olivia or Valéry to walk with her in the park every day, to breathe some fresh air.

Uncle René had just announced that he would show her around Paris personally – the Grande Exposition and the Bois and the Louvre and the Tuileries and the theatres. He was a strange man. She thought him sometimes rather over-critical of his wife, but it was not from any obvious ill-temper. More that he seemed to be a little cold and distant. Always perfectly polite and careful to consult her on small matters and tell her bits of news, but as though it were an obligation, not because he felt she wanted to know. As a married couple they were not at all like Papa and Mama. Aunt Sarah talked much less than Mama, of course, and Uncle René more than Papa. But they did not discuss the sort of things that were a commonplace at Langthwaite where her mother would describe the latest review of some book to her father who would listen gravely and sometimes even ask her own opinion. Mama was a little odd, of course – it was queer she had married a countryman like Papa, but of course Papa was really a businessman and would discuss trade and politics with Mama as well as books. Uncle René was interested in the theatre, he said, but he never discussed it with his wife. Nor did he ever ask her opinion on anything important, she noticed. Aunt Sarah's own opinions you never knew, but she always made an effort to agree with Uncle René's. Yet the strange thing was that Louisa knew that her father was the stronger character of the two men and Aunt Sarah was probably more self-confident in some ways than Mama was. She puzzled a little over this new view of her parents before looking down again to inspect the intricacies of the imperfect subjective. It was much easier and more pleasant to learn French by listening to it. She did not understand all she heard of course – not yet –

but she intended to improve so that at the end of three months she would return understanding it perfectly and speaking it with a better accent.

Olivia came out of the apartment which was on the first two floors of this old *hôtel* with its large black Porte Cochère. Now, if she listened to *her* she would learn something. Of course, Livia spoke English too, but Louisa had implored her not to utter any and to refuse to explain in English. Olivia was not a studious sort of person and neither did she like horses, Louisa discovered. She liked talking about what Louisa eventually realised was furniture and decoration, subjects in which Louisa found little of interest. Still, people themselves were interesting and Louisa determined to try and understand this girl of sixteen with whom she had so little in common. Valéry was nicer, she thought. He was awfully fond of his Mama and both his father and his sister shouted at him a lot. Poor Valéry – he was rather clumsy and was not shining at his school either. He had had to repeat a year because his French was not up to standard. '*J'aime les mathématiques*,' he said shyly on Louisa's asking him his preferences. More than that he would not say, but Louisa was determined to find out.

This particular evening the children's grandmother, old Madame Boissier, was to come to dinner. She was Aunt Sarah's own aunt too, it seemed, for Aunt Sarah's Mama, Lady Gibbs, was half-sister to Madame Suzanne as she was called. Madame Suzanne was a rather querulous old lady, Louisa decided, and she insisted on speaking a hideously accented French, which astonished even Louisa. Sarah, of course, spoke to her in English, but there did not seem any great sympathy between the old lady and her daughter-in-law-cum-niece.

They all sat around in the salon once '*Grandmaman*' or 'Mamie' as her grandchildren called her, had been seated in a tall, high-backed chair, a footstool at her feet, with her son René hovering over her and tucking in a silk shawl around her shoulders. Dinner was to be served at eight, but there was the awkward interval of sipping an apéritif and trying to find topics of conversation. Olivia did not say very much – it was left to Valéry to answer her searching questions on his studies and his plans. Then she turned to Sarah. 'You are looking rather peaky,' she said in English before turning her gaze on to Louisa, who had dropped a curtsey to the old lady and stood around until she was seated. René was now sitting at the window, having done his best to make his mother comfortable. He seemed to have the gift of abstracting himself from the company when he was bored and his mother said nothing, but occasionally turned to look at him. He sensed her look then and always smiled.

'And you are Louisa, yes, Penelope told me about you,' said Mamie, taking out her lorgnette, which was pearl-handled and hung from a silk tassel at her waist. 'Penelope was too good to your Mama – and to Miss Mesure,' she announced.

'Really, Aunt,' said Sarah, but the old lady went on.

'Launching them into society – I heard all about it. Dear me, the expense! It was a good thing my son was invited, for I don't think she had any idea of how to launch her own daughter.'

Sarah coughed and tried not to look cross. Her mother had never once said anything against either Marianne or Kate and all this was a figment of the old lady's imagination. Sarah had realised in the early days of her marriage that René's Mama had always been jealous of her half-sister Penelope and this was her way of getting even with her – but to carry it on for twenty years! And to think that René had once thought to please his Mama by marrying her niece. She knew very well he had, and also knew there was no way of pleasing the woman.

'Maman,' began Sarah, with a quick look at René who was looking out of the window once more, 'Marianne and Kate were – are – my best friends and Kate gave wonderful parties in London at her uncle's. Mama was always pleased that Louisa's Mama (she smiled at Louisa) should have found her husband through our own house parties. You must admit, Mama had great success that year.'

'Well, of course, that American lady found hers too, didn't she, with poor Guy Demaine – such a nice boy, too young to marry.'

'I think he was the best judge of that,' said René, coming forward and pouring another glass of *porto*. 'Really, Maman *chérie*, Guy was most fortunate – they have a wonderful *ménage*, you know.'

His mother was silent and decided to concentrate on Louisa once more. 'Does she look like her Mama, then?' she asked Sarah, as though Louisa was half-witted.

'I think so,' replied Sarah with a smile. 'It's true, you know, Louisa dear, you remind me of Marianne at your age – though I know daughters don't like to be reminded of such things.'

'Well, *your* daughter is certainly not like *you*,' said Mamie Boissier rudely. 'She takes remarkably after me. Come here, child, and let me look at you.'

Louisa breathed a sigh of relief. This was obviously a conversation that was repeated every time the old lady visited. Old people didn't seem to mind being rude or saying the same thing hundreds of times – she had noticed it with her own grandmama, Mary-Ann Amberson. Old ladies were determined to find resemblances between people – and especially that they looked like them or their dear, dead husbands.

Olivia certainly seemed to know what to do for she knelt at her grandmama's feet and there followed a conversation as to the disposition of Madame Boissier's jewellery after her demise. How strange these French were – even Olivia and Valéry. They were always talking about money and dowries, she thought. All the visitors so far had talked about money or possessions or clothes. She didn't think she would want to live in France for ever, though the food and the wine were wonderful and the young women certainly were more chic than at home.

'Olivia est plus grande que vous,' said the old lady, then, turning her gaze once more on poor Louisa.

'Oui, Madame,' replied Louisa, feeling the honour of England was at stake. 'Ma mère est plus petite que Tante Sarah.'

'That's what you call her is it?' Olivia est très élégante, n'est-ce pas?' she continued, patting her granddaughter's hand, which now lay in her own.

In England you were taught that to make personal remarks was rude, Louisa thought. Olivia's grandmama had certainly acquired French characteristics. She decided however to concur. 'Oui,' she replied.

'Olivia est très élégante – elle est française, après tout!' This seemed to mollify Grandmaman and Louisa thought she was casting around to say something about Valéry when the maid announced, 'Le dîner est servi, Madame,' with a frightened glance at the company. There was then the long process of getting Madame Boissier out of her chair and collecting her various bags.

Louisa sat next to Valéry, who was the most like Sarah in character. René said little, sat at the head of the table and passed judgement only on the dishes. Sarah looked flustered. His mother was served first and Valéry whispered, 'Tu sais Maman a invité Tante Kate – à nous rendre visite – elle veut faire ta connaissance.'

Louisa was glad to converse with him for she understood his French better than the others. 'Madame Demaine? Oh, ta mère ne m'a rien dit.'

René had overheard. 'Sarah – I didn't know – is this true? The divine Kate has consented to grace our cuisine bourgeoise?'

'Yes,' said Sarah, busy passing round the plate of hors d'oeuvres. I was going to tell you – I had a letter this morning. Guy is away seeing to their house in Tuscany for the summer. Somehow she'd heard Louisa was here, and wants to meet her. You remember her, of course,' she said, turning to Louisa.

'Oh, yes, of course. I remember when you invited us all to Lake Hall Park, when I was a child, I'd love to see her again.' It was Miles she wanted to see, but she said nothing about that.

'Her son is in the States,' said René. 'In Boston, but I expect he will return when his college semester is at an end.'

'That'll be in June,' said Valéry. 'C'est un chic type, Miles.'

The old lady was too busy struggling with her false teeth to take part in this discussion and Louisa was glad. Miles to come in June! She had so liked him. How would she find him now?

'Miles prefers the States, I think, to France,' said Sarah with a little frown. 'He's a very intelligent young man, Louisa. Of course,' she said to the assembled company, 'Kate is awfully clever too – she is a great success here in Paris. We don't see much of her, Louisa.' Louisa thought Kate sounded terrifying.

'Madame Demaine is very clever with her investments,' remarked the old lady, who had finally managed to extract a piece of *saucisson* from her dental plate and had taken a good gulp of dry white wine which René had poured them all. Louisa loved the wine. It was chilled and yet had a flowery taste like summer distilled in a bottle. She sipped it, thinking of Miles and of Lake Hall Park. 'We all loved your house, and the gardens and the lake and the party you gave. Doesn't it seem long, long ago?' she said to Sarah.

'It's only ten years – that's nothing to us,' said René, smiling for once. 'You were a very *jolie* little girl then and now you are a *jolie* young lady.' He lifted his glass to the rest and solemnly drank.

Sarah was rather embarrassed and his mother didn't look too pleased either. 'Madame Demaine is still a handsome woman,' she remarked. 'Of course, with money you can work miracles.' She looked over at her daughter-in-law as she said this.

Louisa, of course, did not know what a constant source of

annoyance it was to Susan Boissier that her son had not made a great success in the legal profession and the rest of his wife's dowry was all in English land. Why ever he had wanted to marry this large, horsey niece of hers was past her comprehension. She was neither smart nor rich and when she was not away in England, taking her own grandchildren out of France, she was liable to go creeping round churches. Penelope had always been a fool and her husband had neglected to plan ahead for his children. Now they were saddled with Lake Hall Park which ate money and provided little return.

Of course, it was her own husband with whom she was really annoyed. Dead now for twenty years, she had never forgiven him for leaving less than she had expected and some of it entailed to René. Not that she grudged René, her handsome son, having his father's money, but he could have waited to marry and made a good match with a rich daughter of the vineyards. René knew what she was thinking – she had said it often enough – and he had always explained that the phylloxera would have ruined them all if Papa had not sold out when he did. Only now were the new strains of grapes proving resistant to the disease, but it was too late for the Boissier family. To think that he had married when he did partly to please her.

'Yes, Kate is very handsome,' he said to his mother. 'Did you know, Louisa,' he asked, 'that Winterhalter painted Kate Demaine? I don't know where the portrait is now – Guy was so proud of it.'

Louisa had not known and René explained about Winterhalter. René's large brown eyes, so unlike his mother's, rested on Louisa, and Sarah noticed. Years ago René had said to his wife of Louisa, as he had once said to Louisa's mother, 'She'll be a beauty,' and Sarah had agreed. Now she was aware that René looked at many young women in the same way and pronounced upon their looks and general demeanour quite often too. If only he did nothing but look at them, but he was now a confirmed *coureur de femmes* and over the years he had become less discriminating. What had started, she thought, with Bella Dubois, had continued with a whole series of courtesans, youngish women, even very young ones. As he grew older he changed in ways that were more subtle. He used to be discreet but sometimes now it was as though he were daring her to make a scene. He did not bother to invent excuses for being away. She even suspected he had followed one woman to London. It was as though he could not help himself, even that he was engaged in a battle with himself to prove that he was still attractive, had not aged. She was sorry for him but could not let him see that. Her own desires had quietened down and she thanked God for it. Provided he did not land himself in a mess. She took a sip of the wine herself and firmly prevented herself from allowing her thoughts to wander along that path.

'Kate was always striking,' she said. She wondered why René had not married Kate. Money and looks – what ingredient was missing for him not to have succumbed to her charms years ago? The answer must lie in Kate herself.

The delicious meal, with its uncomfortable conversation, wound on from the entrée and the *primeurs* to the gigot of spring lamb and the *tartes aux pommes* and the cheeses almost dripping off their platter, so ripe were they. Sarah ate little, Louisa noticed.

Louisa was thinking of her own father, far away in Westmorland, and comparing him with René Boissier. There was always a specially tender intonation in Aunt Sarah's voice when she spoke of Ned, Louisa felt. She noticed things like that. Father was not a young man – neither was Uncle René, of course, but he seemed younger than her Papa and she wondered why. Obscurely she felt René was a man before he was a husband or a father. Louisa thought of her Papa, suddenly filled with remorse that she had not yet written him a long letter. She hoped he was better – he must take care of himself. Sarah noticed that René's eyes were still on Louisa, who she thought was quite oblivious. But Louisa knew that he paid her attention. It was because he was French and the French all paid attention to women. It was pleasant, but a little embarrassing. She thought, in his case, that he looked at her, not as a woman but as a foreign niece, probably thinking her clothes were dreadful. He was handsome, especially when he went out in his check overcoat with its elegant cape and large buttons and velvet-edged collar, and his moustache was a wonder to behold, but after all, he was Olivia's Papa and probably thought of her as a little girl. She had a lot to learn.

Louisa, however, accomplished a good deal in those first weeks abroad and eventually wrote long, enthusiastic letters to her parents, describing all the places she saw and people she met. It was chiefly the beauty of Paris which seduced her – the long, tree-lined boulevards on the Right Bank, the splendid shopping avenues, the narrow, cobbled, twisty streets that led down to the river on the Left Bank, and the river itself, so unlike the muddy, broad-banked Thames, glittering in the spring sunshine with its ironwork or stone bridges, and the façades of the elegant houses on the Ile St Louis and the towering apartments elsewhere. She was taken round the Latin Quarter by Valéry, who aspired one day to the Faculté des Mathématiques; she was promenaded in the parks with Olivia and her friends and taken to exhibitions of sculpture and paintings and bibelots by René and sometimes Sarah. The churches impressed her – Notre Dame, of course, and others just as old, the Sainte-Chapelle with its glowing windows, and the church nearest to them, Saint-Sulpice, with its square in front, and its fountains.

Everywhere there was light and movement, either in the freshly green trees or at night on the boulevards. Her mother had often spoken to her of the beauty of Paris and she drank it all in. Not that she did not sometimes miss home, though she was not exactly homesick – more place-sick for the hills. Her own vivid beauty and freshness were often remarked upon in the street, though Olivia explained that you must never acknowledge the remarks, but must proceed, head held high and parasol tilted. There were so many elegant women that she was at a loss to understand why anyone should remark upon herself. She gazed surreptitiously at the fashionable ladies on the Champs-Elysées in their dove grey or pale lemon costumes, the bustle less prominent than it was in England, but the embroidery and the fringes and tassels and stitching so incomparably chic. The jackets or the tops of dresses were cut closer to the figure than at home and there were neat cravates and scarves and little waistcoats and sometimes, on the students, soft caps of velvet. Hats amazed her. The fashion was for high-peaked confections which seemed to have begun as a sort of female version of the top hat, but which were then swathed and covered in velvet in bands of tucked ribbons, green roses, pale violet leaves, and which rode on the top of gleaming chignons of hair. These women in their tight bodices and high lace-trimmed collars and with their immense sleeves pinched in below the elbow to reveal a slim wrist with sometimes a bracelet of tiny seed pearls strung on silver, rather terrified her. She would never look like that even if she were rich and French! Her skirts were cut too loosely and made of warm stuff to counteract the cold of the North of England; her dresses seemed clumsy and ill-fitting. In spite of the vast amounts of food which Sarah's staff provided, she found, with amazement, that she had not put on weight. Quite the contrary. 'It's the English food, dear,' said Sarah. 'We have to eat to keep warm at home, and we are so partial to suet puddings and dumplings and roast potatoes and sweetmeats whilst, although I have to serve many more dishes here, they are all lighter, you see.' Louisa noted that Sarah still spoke of England as home.

But more than anything she noticed how fashionably dressed the children were as she watched them bowling their hoops in the Luxembourg Gardens, or sitting in their perambulators with their nurses pushing them along on high wheels. These children did not seem like children somehow – they looked so pert and sophisticated; their clothes never seemed to get dirty and they were aware of them as she had never been as a child. It was a wholly different civilisation and she was torn between admiring it and preferring the more easy-going dress and mannerisms and habits of her own country. Neither did the English seem awfully popular in Paris. They were envied, she imagined, for their industrial wealth, but rather mocked for their way of life, their puritanism and their weather. It was a warm spring that year and she felt as though she were emerging from a chrysalis that had hid for too long in northern climes. But there was no one with whom she could have a really interesting conversation. Uncle René was the nearest, but she was shy of saying anything too personal to him. Aunt Sarah was always amiable, but had different interests, and the children were either wrapped up in *devoirs* or, in Olivia's case, spending hours sewing or drawing or arranging her collection of miniature furniture. It seemed rather childish to Louisa, but Olivia really did seem to know quite a lot about interior decoration.

'Papa once wanted to be an artist,' she confided, 'but, of course, he had to work. I should like to be a maker of furniture or of clothes. Don't tell anyone – it is not a thing *les jeunes filles* can do.' She sighed. She glued paper dolls dressed in the latest fashions which she cut from a special magazine that arrived once a month with the very last word in *la mode*. She had good taste, Louisa was forced to admit; she could always ask Olivia's advice on dress and it would be given dispassionately. 'Papa does not like the idea of me being a "*couturière*," ' she confided, 'but I say what is wrong with a young lady dressmaking?'

Eventually one evening it was arranged that Madame Demaine should visit, and Sarah was busy supervising the polishing of silver and the arrangement of the table. Louisa and Olivia were to be allowed to stay for the *dîner*. But when Kate Demaine arrived with her maid in a grand carriage and Louisa was introduced to her, she did not feel as shy as she had expected. This Kate said rather pointed things and seemed very self-confident, but she was not the fashion-plate Louisa had expected. She wore a simple gown of black velvet and a *rivière de diamants*, it was true, but she was easy and unaffected in manner. Of course, she was American. That must be the explanation. Her conversation was worldly, Louisa supposed, but it was witty and a little cynical. René spoke to her of investments, but it seemed to bore her and she turned to Sarah before dinner to ask about the new horses that were to run that year at Longchamps. Sarah was in her element. Only afterwards did Louisa realise that Kate had done it on purpose to bring her out, for Sarah was usually fairly quiet in company.

Then Kate turned her penetrating, friendly, gaze upon Louisa. It was a light conversation, though Louisa had the impression there were other things of which Kate Demaine wished to speak, but that she was biding her time. She asked about Marianne and Hugh and the little sister. 'Emma, isn't it?'

'Papa has been ill,' said Louisa, 'but Mama thinks the danger is now past. He works too hard, you see.'

'Poor Marianne, she will have been worried,' was all Kate said. Then, 'Tell me, what are your brother and sister like now? I remember them only as children. Do they look like you?'

'Well, Hugh does a bit,' Louisa answered. 'But Emma is taller and quieter.'

Kate laughed. 'You are very like your Mama,' she said. 'How well I remember Marianne – always so enthusiastic about everything.'

As Louisa had scarcely had the opportunity to show enthusiasm about anything yet to Kate, she was a little puzzled. It was always the same. Grown-ups were forever looking for resemblances between yourself and your parents.

'My son is taller than his father now,' confided Kate. 'He is returning from Boston next month. His tastes are very intellectual,

I guess – at least he talks of science and industry and the future of the world a good deal of the time. You must meet him and tell him there are more important things in life.' Later she said, 'I am sorry your father has been ill. My husband is always in such rude health I cannot believe he is forty-seven years old. But Miles says he is old-fashioned. Does your brother tease your father so?'

'Oh, Hugh is a fairly simple person, I think,' replied Louisa, reflecting upon this for the first time. That was the effect Kate had upon you. 'It is I who would have preferred to be a man. It's so difficult for girls to be taken seriously.'

Kate looked at her meditatively then. 'Your dear Mama all over again,' she said.

'Emma is more like Papa,' confided Louisa. 'She is shy, though, and did not want to go to school away from home. She is quite crazy about animals and very good with them. They say she gets that from my paternal grandmother whom I only remember a little. She died several years ago – but she had been an invalid.'

'Yes, I remember – rheumatism. I remember Ned telling us. You must keep your little Emma in the North and let her run a farm. That is what women do in the States, you know.'

'Does your son – Miles – like America?' Louisa finally dared to ask.

'Oh sure – he prefers it to Paris. Yet I believe he likes England best – what he has seen of it, of course. What about you?' Kate, unlike the French, was democratic in her attitude to young people and Louisa knew she wanted an answer. She found herself telling her all about wanting to go to study in Oxford, about her impressions of France and about her opinions in general. 'And what does your father think about all this?' asked Kate.

'Oh, Papa is perfect. He is so kind, he wants me to do what I want, I think. He is all for the education of women, you know – he and Mama are quite agreed. He is unusual, I think. No one else at home – no man anyway but Mr Carmichael – has the same views. Of course, we do not meet the more advanced opinions up in the hills.'

'Well, you must certainly meet Miles again,' said Kate before turning her attention to her other neighbour at table.

Louisa took a draught of Burgundy, feeling quite thirsty after her long conversation. Aunt Kate Demaine was a very interesting person. She wished she could invite her North. Miles, though, sounded even more interesting!

15

When Sarah was worried or tired, she had formed the habit of entering her 'parish church', the impressive Saint-Sulpice. She did this on her way back from the market on the rue Napoléon, for she enjoyed ordering her own vegetables and fruit and cheese and feeling she was a genuine housekeeper and cook. The servants were then sent on later to collect it. It even pleased René to think that his wife was quite capable of knowing about such mundane things; not too grand to put on an old cloak and go out shopping like a *petite bourgeoise*. She was actually much poorer here in Paris than she had ever been in England for René did not take his legal work very seriously. His mother held the pursestrings and was allowed the income from half the money her husband had left, the rest having been laid out in various schemes on the Bourse by René.

Sarah had never left the easy-going Anglican church, which had not yet seen Evangelicalisation, but she was also broadminded enough to consider any Christian church the house of God and to overlook local differences. A compromise had been effected as far as the children were concerned, Olivia having also been baptised into the Anglican Church in Paris that stood near the British Embassy, and Valéry only in the Catholic church, in a smarter quarter than the one they now inhabited and where they had lived at first in a larger apartment. But the Anglican church was too far away to visit without a carriage, and was reserved for Sundays, and nobody stopped Sarah's joining the parishioners, for the most part old ladies, in the great, dark interior of this baroque edifice with its unidentical twin towers, unmatched because one had been left unfinished and now would never be completed.

Sometimes Sarah would light a candle and 'talk' to God as she sat quietly in the gloom. She would never ask for much herself – she scorned that side of prayer – but would attempt to put her thoughts in order, always emerging the better for the process, however inconclusive it had been. The enormous church was often almost empty and an organist often practised high up in the loft above. Music was a comfort. More and more, Sarah felt she needed comfort and was ashamed of her need. Compared with so many of the poor women that she saw in the streets and sometimes even in the gutters, or begging, she was blessed indeed. But one could be comparatively rich and yet unhappy. Sarah was distressed to find that, despite her comfort, she was not in fact happy, or even contented.

It was René, of course, who was the reason for her unhappiness, but she still searched her conscience to discover the blame which she felt sure must be hers. 'I loved my husband,' Sarah said to God. 'I thought he loved me too, and desired me. I still desire him. Is that wrong? When we were first married he thrilled me. Dear God, who put such needs into human bodies, tell me if I should be ashamed at my age of wanting him to go on loving me.'

Before entering the church that morning she had been thinking of her children and of her friends, and of Louisa who had said she would like to go to a Catholic Mass, 'to see what people did there'. Louisa was an invigorating young woman and Sarah liked her. Perhaps she was just too well brought up to say so, but she never seemed bored or at a loose end, which had been the trouble with Olivia till she conceived her present passion for decorations and doll-dressing *à la mode*. Louisa talked more to Valéry than to Olivia. His mother noticed that Valéry had seemed pleased. He was not shy with Louisa the way he was with his father or grandmother. How lovely it would be if in ten years time, say, Ned's daughter could marry her own son! But she tried to stop thinking about this, for Louisa was too old for Valéry and she must not match-make. She began to wonder what Kate's son, Miles Demaine, would be like now. She had not seen him for a year or two and he must have changed, especially after living for a year in America. Miles had not seemed particularly happy in Paris. Maybe he had enough of gadding around and that was why Kate had sent him to the States to re-acclimatise himself.

She was a fish out of water herself in Paris, thought Sarah, coming back to herself just when she had determined to think about something else. She was still scolding herself when she pushed at the sacristy door at the side of the great building. From that elevation the church had the air of a cliff-like fortress. For a moment Sarah wished that she was indeed entering the little church at home, but she forgot this when she sat down in an obscure corner and heard the organist begin to play. Today he was playing something she had not heard before, some new 'emotional' music. Usually it was Bach, which was full of feeling too of course, but a different sort of feeling. This music sounded like a soul in distress, plaintive.

It must be describing unhappiness, in order to conclude in an affirmation of faith. She waited for a happier section, trying to concentrate her thoughts. She said the Anglican General Confession and that comforted her. Why was there 'no health in any human being'? Surely there was much love and health and happiness in God's universe? René did not appear unhappy. Were all men like that? She was sure that Marianne's husband, dear Ned, was not. If only she could have married Ned. Hurriedly she took her thoughts away from that path and concentrated on her Christian duty. She was a good mother. Perhaps she was not a *bad* wife; it was just that she and René were not suited. He made love to her sometimes still, perfunctorily, as though he were brushing his teeth. But she wanted more than that; indeed she thought she had shocked her husband. After all, though, that was what marriage was for: a remedy against fornication, was it not – as the Prayer Book said? She must try not to think of that in church, although it was said often enough at weddings.

Had Aunt Susan had the same difficulties adapting to Parisian life and a French husband, and was that why she was now so sour and bitter? She prayed for Susan and then for her own mother and a sharp picture of the water meadows at Lake Hall Park came before her eyes. Suddenly she heard the music again and sat up to listen. Now it was triumphant and happy – it might be a sign. 'Make me contented, O God,' she murmured and sat with her hand on her chin listening. Other people were now walking up and down the aisles or lighting candles as votive offerings. The candles burned with a flickering strength in the gloom. 'Send me a sign,' prayed Sarah again.

Just at the moment that Sarah was beginning her prayers in Saint-Sulpice, Louisa was walking in the same district. She loved walking alone, though it was not supposed to be done. If accosted she pretended not to understand a word (and usually did not). She saw long, narrow streets with tall houses of faded ochre, beige, fawn, yellow, dirty white; then a square studded with cobblestones; chestnut trees, each growing carefully from a circular protective grating, the trees now the gay green of summer. In the square was a great four-sided fountain with niches for statues of the forgotten famous, guarded by four lions couchant at each corner and topped by a lacy baroque stone canopy, water eternally tumbling in small Niagaras over its shallow steps to a pool below. And she lifted her eyes to the Square's *raison d'être*, the grandiose St-Sulpice: a monument, thought Louisa, to past glories, but she was not sure enough of her French history to know which glories exactly.

Louisa walked up to the church, and pushed open a padded inner door, wondering whether she would have to declare that she was a Roman Catholic before she was allowed in to the arching gloom. Clusters of candles, like haloed chrysanthemums, flared in waxy, dusty corners; stone or plaster saints, each in his or her scooped-out upright bed, brooded over the side chapels, and in the nave the *prie-dieux* were ranged neatly for the next Mass. The pulpit reared from one of the centre pillars like the home of a looming bird of prey, gilded not gold.

She saw the sanctuary light glowing in its artificial blood-like scarlet, awaiting a curtsey which she hesitated to make, feeling more foreign than ever before, excluded from a community. She walked round quietly, passing a woman here and there at her devotions. She paused near the end of one flagstoned walk and looked over to the nearest little altar of candles, thinking she might light one and wing a prayer home for Ned's recovery. Then on approaching more closely she was suddenly stopped in her tracks. Was not that Aunt Sarah, of all people, sitting near the candle-lit statue, her eyes closed? Then she realised that an organ was being played high above, very softly. Louisa sat down some distance away to listen. It sounded now like music with which she was familiar, yet something rather new, music she had listened to in Manchester with her mother that very winter. It was certainly French. Louisa closed her eyes for a moment, but found prayer did not come. Should she just tiptoe away? Sarah might be embarrassed to be found praying in a Catholic church, for it did seem as though that was what she was doing. But just when she had decided to move, Sarah got up, bent down for her reticule, and then turned. Louisa thought she would pretend to have just come in after hearing the music. She advanced vaguely towards the nave, pretending she had not seen her. But Sarah was coming towards her now with a smile of welcome.

'Louisa, dear – how nice,' she whispered, adding, 'I often come in here – it's so peaceful.'

'Yes,' replied Louisa.

Sarah looked calm and a little sleepy. She was thinking, 'This is my sign – this girl, dear Ned's daughter.'

They went out together into the square and walked slowly home the long way round by the gold-tipped railings of the Luxembourg Gardens. Sarah took Louisa's arm.

'Let's just say we've been for a walk, shall we?' she suggested.

'That's right,' answered the younger woman. Then, 'Mama likes churches too, even though she no longer takes the sacrament – and I was not confirmed.'

Sarah pondered this. Marianne had always had doubts, which were probably shared by her husband. But Louisa must still be searching – unless she just liked exploring. She said nothing further on the subject.

Miles Demaine was to return from Boston in June. Louisa looked forward to meeting this paragon whom she remembered very well from ten years before, although she had forgotten much of that holiday at Lake Hall Park. Forgotten all except Miles and the shining Christmas tree – and a vaguer memory of roller skates. She thought of him as a paragon since Sarah seemed to like him so much. What did she know about him though? She went over what the others had said, so that she would have some conversation when he finally made his appearance. He preferred England to either France or the States. She wondered whether she did too. Probably. After all, it was home, though France was superior in many respects. English countryside had, as yet, no rivals. It was not quite fair – she had not been to the French Alps or to other parts of France, had seen only the fields of Normandy on the way and the lush pastures of the Ile de France, though they had had a day out at Bougival and had also been to the races which had rather bored her, though she had tried to show interest for Sarah's sake. She hoped that she and Miles would share similar tastes, was determined they would, though why she should wish to was beyond her. She had nothing much in common with Olivia, and Valéry, nice though he was, had no real conversation. She would, however, like a young man for a friend, she decided. Someone to talk to.

When he was finally announced – he had come along alone one afternoon, though whether at Kate's insistence or not she did not know – she was surprised. He was tall and dark and really handsome. But he had a free, frank air about him – American, she supposed. He was not at all shy and shook hands in the salon with herself and Olivia and Valéry, Sarah officiating. Valéry greeted him like a long lost friend and seemed overjoyed. He had some intention of discussing photography, Louisa gathered, till Sarah said gently, 'Time for that later, dear – you must tell us all about Boston.'

Miles, who had smiled at Sarah and then sat casually in an armchair, having accepted a glass of Barsac and a piece of cake, said, 'Oh, it's just fine – lots of concerts and lectures. We had an exeat one evening a week and I went with my mother's cousins to all sorts of stuff – scientific talks and theosophical societies, and even to a meeting for Women's Suffrage. Boston is still in the vanguard, you know. I had to work pretty hard too – my Greek was not up to scratch.' Louisa thought him very modest and very agreeable and wanted to find out more about the Suffrage meetings. 'There are all sorts of modern notions over there – the girls are much freer than in Europe,' he added.

'So, you're going to return to Europe for another year?' asked Sarah.

'Probably, I still haven't decided what to study. Mama wants me to study law and Papa thinks I ought to become a businessman and make a lot of money,' he laughed.

'But what would *you* like to do?' asked Louisa.

He turned his serious grey eyes upon her. 'I don't know. I'd like to be a scientist, I think – or a farmer. The Americans are catching us Europeans up, you know – industry and farming on a vast scale, and so many new patents. I must show you the Camera Obscura game, Val,' he added. 'I have a camera of my own my great-uncle gave me.' He talked on, full of enthusiasm, but always stopping to ask a question of Valéry or an opinion of Sarah. Louisa noticed that he did not say 'we' of the Americans, but 'they', so presumed he did not regard himself as one. Sarah thought, Miles should be unsettled with all the upheavals he's gone through, but he's not. He looks happy now.

'And how is your Mama?' Sarah asked.

'Fine – she told me about coming over to see you a few weeks ago.' He turned to Louisa. 'Do you remember the Rinkomania?' he asked, laughing.

'The Rollerama, you mean,' said Louisa.

'Of course!' Sarah exclaimed. 'Fancy your remembering that.' Then, 'You must excuse me – I have some work to do, Miles. I'll leave you.' She went out with a smile. Olivia went after her so Louisa was left with the two men one fifteen, awkward and shy, the other just about the nicest young man with the easiest manners she had met in her life. He was what her Papa called an intellectual, she could see that. Suddenly she wanted to ask him about his own views on God, England, Feminism and John Ruskin. How could she begin?

'I brought you some stamps,' said Miles to Valéry and fished out an envelope bursting with American stamps. Valéry, who collected them, was thrilled and ran to fetch his album. 'What do you collect?' asked Miles of Louisa.

'Nothing really – I used to collect fossils and pieces of rock, but I was too lazy and disorganised, I'm afraid. Were there many Feminists in Boston?' she asked before he could reply.

'I guess so – are you a follower of Annie Keary and company?'

'Of course – Mama worships at the shrine of a lady in Manchester called Becher, and reads the works of Swanswick, Webster and Winkworth. We know the Edwards sisters too – they're quite famous in the North of England.'

'I forgot, you live in the North. What are *you* going to do next?' he asked politely.

'I want to go to one of the ladies' halls in Oxford – '

'To study what?' He looked a little more guarded. Perhaps he had thought she had no brains.

'Well, French, I suppose – that's one of the reasons I'm here!'

'And how is it, your French?'

'You will have to tell me for I don't know. We often speak English with Aunt Sarah – I expect you speak French and Italian as well as you do English?'

'Not really, I feel more at home with English. My Mama never spoke French to me, always English. Of course I speak French sometimes with Papa.'

'Aunt Sarah was saying – or was it your Mama? – anyway, that you prefer England to France? But you haven't been in England much, have you?'

'A little. I liked what I saw of it – London is too big, I think, but it's comfortable. Of course England itself is small – and you feel it will never change, though it should.'

'What else do you remember of Lake Hall Park?' she was emboldened to ask. 'As well as the Rinkomania?'

'The Christmas tree,' he answered. 'And you – you were very talkative. I do remember that.' He laughed.

'It was a happy Christmas. That tree, it was so beautiful,' she said.

He looked at her and saw the faraway look in her eyes. 'Are you interested in science?' he asked, changing the subject.

'Well, interested – but very ignorant. I have a great-uncle, Mama's uncle, who is a scientist, Mama's favourite relative. But I don't know where to begin. Mama tried to read *The Origin of Species* and got stuck in the middle. Girls always botanise,' she went on. 'But we are taught very little of geology and natural philosophy.'

'Oh, you should read Manson,' he said quickly. And went on to talk of other books she would enjoy.

Louisa cleverly got him on to music and then to novels as she wished to shine a little and he listened to all she said. Yes, he had read *The Arabian Nights* in Burton's translation and Henry James's *Bostonians* – it was much more popular in England than Boston, he added. They had both read *Jekyll and Hyde*, but only Louisa had read *The Mayor of Casterbridge*, which he really must read if he liked England – and Meredith too. 'Ah!' said Miles, '*Diana of the Crossways* – a feminist book, I think.' But he had not read Walter Pater and Louisa promised to lend him her *Marius* which she had brought to Paris with her. 'Of course I read lighter novels too,' Louisa confessed, and they discovered they both had a passion for *King Solomon's Mines*.

The time passed so agreeably that they could not believe it was six o'clock when Sarah came in again to ask Miles to stay to supper. René would not be with them, but they would love to have Miles with them a little longer. Louisa had never had such a long conversation with a boy of her own age. He had a most interesting mind, this Miles Demaine, and she would love to be his friend, to be treated like a '*copain*' of his. More than any other young man he actually took her seriously. It was blissful. Perhaps they could be real soul mates, she thought. Nothing spoony or silly, but really close friends. If only he lived in England.

She lent him her *Marius the Epicurean* and wondered when she would get it back.

Louisa was not, in fact, to meet Miles again for some time. He had to accompany his mother to Italy to shut up their house in Tuscany and go shopping on return to Paris, a mammoth affair. Kate had a new house now to furnish; everything must be perfect before she would entertain in it: curtains, carpets, furniture from auctions of past masterpieces, bibelots of all kinds, but always in the best taste, and a new staff to employ. She had decided to make another effort to ingratiate herself with the *crème de la crème* of Parisian society, but at the back of her mind now she thought of returning one day to the States and perhaps being a much larger fish there with her Continental manners and an air of fashionable chic. She had been a successful hostess in Paris up to a point, and in Florence and Rome, but she had never felt herself exactly accepted by the French, and the Italians, charming as they were, did not count. Guy was against any move across the Atlantic, however far away its date might be, so Kate gave him his head over tapestries and furniture and kept an eye on his gambling. One thing that America would do would be to wean him away from the gambling tables of Europe, but now was not quite the time. First of all they would make another splash with the fortune that had recently accrued from shrewd railway investments in America. The west coast was being opened up and there was even more money to be made.

Kate's daughter, Kitty, the reputed beauty, was like her father in her artistic pretensions and her charm. Although that spring she was only just fifteen, she looked and acted like a much older girl and Kate meant to secure a prince for her before going to the States. But if the gamble of a new apartment, new parties and expensive accoutrements did not pay off, there would always be a rich American. As yet Kitty seemed pleased to follow her mother, whom she admired, and to be a companion to her father, whom she adored. Kitty's only weakness was a penchant for clothes. She had the largest shawl collection of any of her girlfriends. She loved clothes with a passion that even her well-dressed and handsome

Mama did not quite share. Already she had been painted as *Jeune fille a l'étole* by a fashionable Parisian painter, and already lascivious looks had been cast upon her by Italian youths. This was one reason for a return to Paris where manners were more formal. Louisa had not yet met Kitty who always seemed to be away somewhere. Of course Kitty, as Sarah said, would not be out for two years, but protocol on the Continent was not quite as severe as in some circles in England. Girls married in Paris sometimes as young as sixteen.

One morning a letter arrived for Louisa with an Italian franking. From Miles Demaine. Her first letter from a young man. But Louisa did not think of him as a beau at all and was completely unselfconscious when she opened it. He had told her on their long, happy afternoon that he kept a journal too. The letter described the countryside around Florence with its poor peasants and rich landowners, and mentioned that he had been practising the art of photography when not helping his Mama to pack. 'The maids are supposed to do the packing, but Mama is very particular and oversees everything. It is too hot here. I long for green meadows. How long are you staying in Paris? Paris is dead in August, full of tourists. I believe Uncle René usually goes to Bordeaux with his Mama. Will Aunt Sarah go on later to England? If she does you might mention that I should love to see Lake Hall Park again!' After discussing his present reading, which seemed to Louisa to be an alarming mixture of political science and history, he went on: 'Do you know the Fabian Society in England? They are very much of the right way of thinking to my mind – you must read their prospectus. They call themselves the Fellowship of New Life, grand isn't it?' He continued with a description of a concert he had attended with Kate in the city. 'Such operatic arias! You really must hear Verdi – or perhaps you already know his work?'

He was intelligent and very old for his age, she thought. Or perhaps that was just when you compared him with the young English lads of her acquaintance who were often tongue-tied, or talked only of sport. At last she would be able to have a correspondence where she could pour out her own feelings. She took pen to paper immediately, her head buzzing with ideas and notions. Notions she would not have spoken about to anyone else, even her girlfriends Jessie or Ethel from school who talked about clothes and curates, only occasionally about religion and families and friendship. Miles was a change.

Sarah was thinking that Marianne and Ned's daughter was what the French called a *bas bleu*. Like her mother she was, in so many ways. Louisa told her a little of the contents of her letter from Miles. He preferred intelligent girls, thought Sarah. She hoped he was an honourable young man. But with Kate as a mother . . . Sarah stopped her uncharitable thoughts. Why should she have anything against Kate, who was always charming and had, in fact, been particularly so when she visited them? Miles was an unusual person. She'd thought when he was only a child that he had great intelligence and imagination. Not that she understood all that Marianne used to talk about as far as 'imagination' was concerned. But she was pleased for Louisa's sake and a little spark of future possibilities was fanned in her kind heart.

'Are you set for Oxford?' she asked her young guest that same evening.

Louisa had been writing all day and had then gone into the courtyard with a book. She came in glowing with sun and health.

'I do want to be educated, Aunt – I only hope that I am clever enough and that Papa will be able to afford it. I don't intend to get married – or not for years and years!' she added.

Sarah smiled to herself. 'It is so difficult to get to know young men as friends,' Louisa went on ingenuously. 'Of course, you see, I am a Feminist, Aunt Sarah . . . the world is *so* unfair.'

'But wouldn't you prefer to marry and have children?' asked Sarah. She supposed that Louisa must have this sort of conversation with her Mama, for Marianne had always been one for discussion of Big Subjects.

'I don't know. When I think of being an old lady and never having any descendants, I suppose I would like to one day. But you see marriage is slavery, isn't it, for most women?' Then she stopped, aware that she might have been rude. 'Papa is not a slave-driver,' she added. 'Mama always says she is lucky. Mama has a friend in Kendal, a Mrs Warburton, who has eleven children – it is quite common, you know.'

Sarah blushed. She wondered how much Louisa knew of women's intimate lives.

'I should like just two or three,' she went on. 'Though I'm afraid I don't show enough interest in my brother and sister. Emma is the sort of girl who will have lots of children and Hugh – well, it won't matter for him because he's a boy. Oh, don't you wish you'd been born a boy, Aunt? I do. Have you read *Little Women*? Well, I've always felt like Jo. Do you think America is better for girls?'

Sarah, who knew nothing of America but did know her friend Kate, was not too sure. Kate had never given the impression of being emancipated from the duties of a woman in society. 'I haven't read it – I don't read a great deal, dear.' Then, 'Don't you think it pleasanter to be a woman, Lu? Men have to work so hard and have all the responsibilities, earning money for their families. Women can have more time to concentrate on people.'

'But *I* should like to work!' cried Louisa. 'I don't find collecting shawls and dressing up and flirting very interesting, you know. And it's so much more practical to dress as men do, you must agree. All these skirts and petticoats and frills and ribbons – ' She looked like a rebellious child. 'I hate going to dressmakers – such a waste of time. Of course, I do like pretty things and I think Olivia is very clever – she makes an art of it, doesn't she? – but I'm hopeless at all the practical things women are supposed to be good at. I'd like to learn Greek – we didn't at my school.'

Sarah was thinking in a puzzled way that Marianne and her husband must really have joined what her own mother called the Middle Classes. Ned had never been a snob and she gathered his way of life was simple and hard working. 'Well, both your mother and your father came from business people,' she said tentatively. 'Provided one has enough money I can't see why one may not live however one wants – but women's lives are always different from men's.'

'I know, but that's because women do what men want them to, you see,' replied Louisa with a flushed face and sparkling eyes.

It was as well, thought Sarah, that some young men liked to correspond with independent girls. She could not imagine that René would accept it with good grace if *his* daughter started discussing women's rights or balking at her domestic domain. Yet really she herself would have liked to be a boy when she was young. Then she could have taken over her own papa's estates.

'I think Miles is awfully nice,' said Louisa though this did not seem to have any connection with the previous conversation. 'He's intelligent and honest, isn't he?'

'His Mama always says he has an obstinate streak,' replied Sarah.

'I think he is a good, upright boy.' She left it at that. Louisa went back to her room to add some new thoughts to her letter.

Sarah was left wishing that she had had a confidante, as Louisa must have in her own mother. Lady Gibbs had never encouraged confidences, though she had always been a kind woman, and her mother-in-law, Tante Suzanne – well one could hardly expect *her* to have a woman's interests at heart when she was so devoted to her son. Did she know that her son was running through a second generation of women now? And if she did know, would she care? Sarah wished that she could have a nice long talk to Marianne. There were some things you could not say to young women about marriage. She would write to her instead.

Louisa was to return to England at the end of July, so was careful to send her English address to Miles when she finally posted her missive, in case he did not reply till later that summer. Then she began to read a silly novel that Sarah had lent her by a lady called Marie Corelli, but threw it down quickly in bored disgust. She decided that she would spend the next morning keeping up with her other correspondence. She loved writing letters, especially to her father. She felt that she worried him a little sometimes. Had her mother been impetuous and excitable like her? Ned was always telling his wife to 'take things easy'.

The evening that Louisa's letter arrived home in Langthwaite, Ned had been up and for the first time after his long illness able to go outside. But he was feeling better every day and was determined to return to the Manchester factory in a week or so. Marianne had been having a hard time nursing him. She was not cut out for it, worried too much and then felt guilty that she wanted to escape the house and her domestic obligations and go for a walk round the lake or take her sketching pad into the woods. That day had been long and sunny and he had urged her to get out and inhale some fresh air. 'I can't have you missing your summer,' he had said. 'Polly and Mrs Longbottom will give me my lunch and Emma and Hugh are perfectly happy on their ponies. You go for a little walk. I'm not quite up to it yet.'

Marianne felt released from a burden as soon as she escaped the house and the garden, which always needed something done to them and had been neglected while she nursed Ned. Once out and in the woods that circled the lake at this point she felt young again.

The countryside, tamed down here but grand and wild in the distance, soothed her in the way that, of her other pleasures, only music could; in it she felt more in touch with some essential Marianne whose presence she felt only when alone. The hills reduced her to size and the ever-changing skies hypnotised her. They made her feel both serene and piercingly happy, though happiness was not quite the right word. In touch with something, would be better, though whether it was a buried self or a release from self she was never quite sure. She called it 'reality'. She cherished these times that never came when they were called, but arrived involuntarily with the scent of the conifers and the ferns and the low murmur of the wind. It was as if Nature was saying it would all go on when she was gone and she did not mind. Yet this place with all its beauties could also overwhelm her sometimes. Too many daffodils, too many distant pastures, lakes, meres, forests. If she calmed down and concentrated on just the one flower and the one lake and the one tree she could feel at peace once more, and surrender to the special feeling that went with a delighted vision of hundreds of pale pink rambler roses that tumbled and swarmed over the hedges in scented splendour; or the cream-coloured cloudy veils that were spread round a full moon, shining over the lake. Or, today, when the clouds were different again, like wool teased by a woolcomber to broad yet fluffy strands sweeping over the arch of blue.

She was thinking as she sat down under an oak tree, I must not think too much about my own feelings. I must be thankful for my life. Nature has done her healing work on me. Natural beauty is not a trap, as human beauty can be. She turned her thoughts away from human beings and looked up at the dancing sunbeams crisscrossing the branches above her head. This beech wood was beautiful as were the larches further along and the paths that wound through the woods with the sun occasionally descending in little pools on to the leafy ground.

She belonged now to the landscape of her married home. Once she had felt that to be contented she must somehow master it, understand it, sketch it, but now she was content to belong to it. Her home gave her happiness also in winter: firelight on jewels; old polished furniture, her comfortable feather mattress, the smell of bread baking in the kitchen. Just now she sank into a warm, light confusion of sensation – the cowbells in the distance, the bleat of a sheep, the feel of the sun through her simple cotton dress and the sight of the purple vetch that twinkled under the hedge. She closed her eyes and was, for an hour or two, supremely happy, not asleep, but poised between sleep and waking in a delightful land.

It was, then, a double shock for Marianne that, on her return to the house, Ned should exclaim to her as she entered the drawing-room, 'A letter from little Louisa, Marianne. Such a long letter – I'll read it to you.' And that the letter should contain a long description of Miles Demaine, the only person she had not wanted her daughter to meet. But had she not known it would happen? Had there not been some fatalistic disregard of her fears that had allowed Louisa to travel and thus to meet her half-brother again? Now she must gather together her forces and make a stand. Perhaps there would be nothing in it, but she had better tell Ned, as diplomatically as possible, without letting him guess anything untoward, that she would disapprove of a match between the two. A match! What was she thinking? Louisa only wrote of him as an intelligent young person with whom she had had a conversation more interesting than was usual with young men. And she knew her own daughter well. The slightest hint of displeasure and Louisa would be up in arms. No, she must play her cards coolly. Begin with her husband.

'He sounds a pleasant young man,' she remarked. Ned had finished Louisa's description of a letter in which Miles wrote about 'Feminism and Evolution!' 'Amazing that Kate should produce a Radical,' she went on. Ned grunted. 'Of course, Louisa must meet young men, I suppose, though I must say she has a lot of work to do for Oxford and mustn't get her head turned by compliments as to her intellectual abilities.'

'I expect he's a charmer like his father was,' Ned said, rather sharply.

She looked away. 'Kate was quite clever, of course,' she replied after a pause. 'But I wouldn't want Louisa getting mixed up with a Catholic.'

'Oh, he sounds like an unbeliever to me,' laughed Ned.

'Of course, she's far too young to think of any young man *seriously*,' Marianne went on.

'Of course,' replied Ned. 'She's only a baby – I expect she missed decent conversation at Sarah's'.

'Yes, I suppose so. Still, we mustn't encourage too great an intellectual intimacy on Lu's part with *any* young man yet.'

'I'm sure there's nothing in it,' said Ned comfortably. 'I'd be the first to agree with you – Louisa must finish her education first.'

'Oh, yes. I just thought it's the first time she's shown any interest in male conversation.'

'There'll be hundreds of young men later,' said Ned.

'Chaperonage will be strict in Oxford,' mused Marianne.

'Listen, she has something to say about René,' said Ned.

He read Louisa's paragraph on René's conversations with her: ' "He praised me up to the skies – really I'm getting quite spoilt with all this praise. But Uncle René and Aunt Sarah are like two civilisations clashing. You know, I don't think I belong to either of them. That was why it was so nice to talk to Miles. Aunt Kate is, of course, quite terrifyingly elegant" – I won't read all that,' said Ned, putting the letter down, then suddenly taking it up again and saying, 'Where was I? – oh – "I didn't meet Uncle Guy this time. I expect he is just as fashion-platey as his wife".'

'True,' thought Marianne. Louisa's reactions were always swift and amusing. 'Does she say anything about her studies?' she asked, to change the subject. How happy she had been in the forest looking at the flowers, away from human beings and their antics. If Ned were ever to suspect, even now . . . and even if he did not, she could not be happy again until the threat of Miles Demaine was for ever removed from her daughter's life. How, oh, how to go about it? As Ned read: ' "I sometimes feel I'd like to be a governess or a proper teacher and teach girls things I've learned myself, except that they have such dreary lives. But I could always do it after my Oxford studies – Papa never needs to worry about my finding a rich husband for I intend to work for my living." ' Marianne smiled. 'Doesn't sound as though she's feeling romantic about the Demaine boy, Marianne.'

'No – I expect she was more concerned to show off her learning to him,' said Marianne, rather unkindly. The danger for the moment, if it had been a danger, seemed with the rest of the letter to be past, especially when Ned said, a few minutes later, 'I'd rather she married an English boy one day, you know.'

'Not a Catholic anyway,' said Marianne in reply, and remembered her own mother's similar words all those years ago.

But Louisa was more sensible, more secure, and had the prospect of serious studies before her, rather than a season in society. She wasn't the sort of girl who thought that life began with marriage as so many young girls did. Marianne decided even so that Louisa's nervous energy must be directed to things other than young men. But could you go against Nature? Would not Louisa want passion one day and find that her studies and her ideas of feminism and the rights of women conflicted with falling in love? Feminism and reading were all very well, but they could so easily lead a girl astray.

16

Louisa had finally returned to England and was on holiday on the Lancashire coast with her mother and sister when Miles returned to Paris. He had expected and hoped to find her still in France, for he had been much impressed by the girl. Unlike her, he had met plenty of the opposite sex and was therefore not chiefly aware of any ease of conversation that she possessed, but rather by her personality and her manner. In Boston bright girls were two a dime, but not girls with Louisa's freshness and looks. American girls were more forward, it was true, than European girls of the same age, but not one of them had stood out from the crowd in Louisa's way. Besides, he remembered her very well from his childhood. He had said nothing much to his mother about her, beyond stating that they had passed a pleasant afternoon together. But he decided to write to her in England and perhaps, if Aunt Sarah could arrange it, one day they might meet again at Lake Hall Park. But Sarah was not yet home in England, and Miles was a young man who spent a good deal of the time studying because he enjoyed it, so he put the letter off for a week or two. Yet he often thought of her.

Louisa herself said little to her parents, beyond confirming that Miles Demaine was indeed a nice young man with whom one might have an intelligent conversation, and for the moment her mother left it at that. She had taken up her language studies willingly, determined to succeed, but she did find herself occasionally dwelling on Miles when she had thought her mind was full of Latin conjugations or the sequence of tenses. Marianne had procured an elderly clergyman to teach her daughter Latin and this Louisa began on their return from Silverdale. As she bent her head over Virgil and painfully translated the death of Queen Dido, Louisa found her thoughts in Paris. But she put her daydreaming down to the lack of company at home, apart from her family. She seldom saw her old friends from the Manchester school now. Two of them were also working for entrance to Oxford and being both daughters of clergy had ready-made tutors at home in the shape of their fathers.

Ned seemed to have recovered from his pneumonia and was back at work, even busier than usual, trying to catch up with all he had missed during his illness.

Marianne went once a week that winter to her concerts, staying with Ned at a small hotel in Manchester on the Saturday night before coming home on the Sunday to Langthwaite. She always returned happy, trying to remember the music she had enjoyed. Brahms and Franck were her favourites now and one day she was humming a tune at home when Louisa heard and exclaimed, 'That's the music the organist played in Saint-Sulpice, Mama. Aunt Sarah enjoyed it.'

'I didn't know Sarah was musical,' said Marianne.

'Oh, I think she often goes to listen to the organ music there,' replied her daughter. 'At least she was probably praying one time we met by chance in the church. Was she always religious?'

'Yes, Sarah was always a good woman,' said Marianne, looking up from the repair of Emma's 'Guernsey' jumper, for she dressed her almost grown-up children in modern fashions and was beginning to favour simpler dresses for herself too. Darning was something Polly could have done better than her mistress, but Marianne liked an occasionally self-imposed task. It made her feel virtuous.

'Did you see the vicar's daughters in church last week?' she asked. 'Both of them in Kate Greenaway dresses – so pretty.' Marianne and Ned went to church because it was expected of them as members of their little community.

'I didn't notice,' said Louisa truthfully.

'Well, you said that French children were so much more fashionable than ours – I wondered if that fashion had caught on in Paris.'

'No, Mama, I don't think so. Do you think ardent or soulful is a better translation of *fervidus*?'

'Depends on the context – what about "impassioned"?' replied Marianne, threading her needle.

'Oh, Mama, that's just right! Why didn't *you* go on studying Latin? Aunt Sarah said you were a bluestocking like me.'

'Oh, did she? My parents had no desire to see me as a *bas bleu*.

Latin was strictly for my brothers and even then they didn't see the point. I had to learn it clandestinely. Do you miss Paris?'

'Sometimes – I loved it – but I think after all I do like England best.' Louisa was going to say 'except for the food', but thought her mother might take that amiss. The cooking in Westmorland was plain but wholesome and Marianne had enough to do getting the elderly cook to try out something other than dumplings and puddings without exhorting her to more aristocratic culinary efforts.

'I suppose it must seem dull here?' Marianne tried again.

'I don't mind – there's always walks. And I'm sure Oxford will not be dull.' Louisa returned to her little bedroom where she had a card table next to the window and her pile of books and papers. Marianne thought all was well for the time being.

Louisa would occasionally pause and look out of the window over at the hills which in winter were usually mist-topped. *Did* she miss Paris? She was not sure. Perhaps she could get on better there as a person because she was a ready conversationalist, but she did not feel she had really enough wit to pass herself off as French. She fell to wondering whether a quality such as wit or a ready tongue was inherited or acquired. Like all interesting questions there was no way you could find out. Miles seemed more American than French, but he had been brought up by an American mother. She wondered what Uncle Guy was like, wished she had met him again.

Since returning home Louisa had noticed things about her parents that had previously escaped her attention. Mama was getting more and more 'mystical', she thought. If it were not for the visits to Manchester she would be quite a hermit. Louisa herself wanted to travel and try her wings, would not be content with country existence. But perhaps you grew less adventurous as you grew older? She had always been at ease with people, and her Parisian holiday had added to her self-confidence. Oxford, though, would be different. There everyone would be brilliant! She had no illusions about her brains, knew she was intelligent, but knew also her own limitations. Still, she did wish to be tried in the fire of academic life. It would be cowardice not to use the small amount of emancipation offered. It was good of her father not to quibble over her further education. So many Papas would have done.

Life should not consist only of work – surely in Oxford even in spite of the chaperonage she would meet many young people and even older ones from whom she might learn something? One could not know of the world only through books. Papa had travelled, although Mama had not been abroad very much. What would Mama have been like if she had never married, she wondered? Would she have liked her if they had been of the same age? Mama was so unlike Aunt Sarah and Aunt Kate – how had they got along together? People were mysterious, no mistake.

Her Mama was at that moment reading a letter from Sarah who had not been able to visit Lake Hall Park that year. Sarah was hopeless at hiding things, or perhaps she no longer wished to hide them. Marianne was once more troubled. Louisa had not said much about René except in connection with herself. Had she noticed that Sarah was unhappy? No, young people were too busy thinking about themselves! But Marianne guessed all that was not said in her friend's letter though perhaps her vivid imagination went a little too far. Things, of course, were done differently in France, as she had good reason to know. People there were not so censorious, even about prostitution. She could easily imagine that for some young women prostitution could, if chosen freely, be enjoyed, especially if they had a rich patron or two. But that was the sort of thought she could not voice in dear old England. The trouble with England was paradoxically that pleasure had to be reserved for the marriage bed, without, of course, mentioning the act, where it might be absent.

Perhaps her views of France were out of date? But even Ned had let slip once or twice that business acquaintances of his took their pleasures in Paris when out of reach of stuffy Albion. She knew that Ned did not, understanding her husband well enough. He did not even visit the cigar divans in London, though he had mentioned with a sort of wondering amazement that quite fresh young girls could be found there, according to his friend Wilkie. Marianne realised her husband was unusual in telling her of such things – most men would not dream of such candour with their wives. But she could see that he was very shocked. It made it even harder to bear that she had sinned herself long ago and had done the unforgiveable, passing off another man's child as his. He would refuse to believe her, she knew, if she ever had the courage to confess. But what good would confession do? It would only make him wildly unhappy. The only problem was between her conscience and herself and she could not think of any way of resolving it short of forbidding Louisa ever to see Miles again, which might only arouse Ned's suspicions. Marianne found herself thinking that if it had not been for Louisa her memories of Guy would now have resolved themselves into happy ones. She *had* 'sinned', but perhaps she was not a good woman in any case. Marianne tried to acquit herself of charges of hypocrisy, but the only other way of looking at it was to say that the congress between the sexes outside marriage could be a good thing or a bad, depending on circumstances; and this rather frightened her in the same way that other people's judgements frightened her. Was there a conspiracy to deny certain facts? It seemed so.

She wrote back to Sarah, whose letter she read and reread. There was nothing said explicitly against René in it; it was almost unconsciously pathetic. Sarah was lonely; she missed England and poured out her tenderness on her children and animals. She hoped to visit England next autumn, but she was at present needed to help nurse her mother-in-law who was suddenly bedridden and required companionship. By the time Sarah was at Lake Hall Park, Louisa would, *deo volente*, be in Oxford. Marianne let matters rest once more and concentrated on being a companion to her Ned, who seemed to have recovered his health, except that he tired more easily.

Louisa took a special Responsions and an examination for Somerville Hall in the spring and was overjoyed to be accepted for the autumn term of 1887 to read French. She was then on holiday with her family once more in Silverdale, a village which she had known since she was a child. Ned had been accustomed to rent a large house almost on the bay for six weeks and thither all three children and Polly would repair with Marianne, whilst Ned came only for weekends and for one week in August. It was a beautiful spot and once Louisa was sure of her future – for the next year at any rate (whether she would pass her first examination at the end of it was another matter which she rather dreaded) – she again began to think of Miles Demaine. He might visit Lake Hall Park in the autumn with Aunt Sarah, she thought, unless he had gone back to Boston again. She had had one letter from Boston in reply to hers the year before and Miles had promised he would see her before the year was out. 'To continue our discussions of the fate of womankind'! If he visited Oxford they would surely allow her to meet a 'cousin', even if she had to endure a chaperon!

That summer her brother Hugh grew into a tall young man, looking more and more like his father, but with a more irascible temper. Emma remained her old self, kind and merry, so long as she was allowed to follow her own pursuits. Louisa consented to trawl the grassy pools of Morecambe Bay with her for marine life and algae and they also made a collection of pebbles and went for long walks in the lovely woods of the area. Here Mrs Gaskell had stayed – her house had often been pointed out to them and Louisa read *Ruth* and *North and South*, which she much enjoyed.

She felt herself to be a child again when she waded out into the bay with her skirts tucked up while her mother enjoyed herself sketching and dreaming. Altogether it was a satisfactory summer for them all.

That year was important in the wider history of the realm, but to Louisa Mortimer all that mattered was her escape, as she saw it, to Oxford, to a wider life, with the delicious anticipation of a further meeting of souls. But Louisa did register the Golden Jubilee of Queen Victoria, although without much interest. Victoria had been on the throne for ever, it seemed – she had even been there ten years before her own Mama was born! She read in the newspaper her father took of the so-called 'Bloody Sunday' when Trades Union activists gathered in Trafalgar Square within sight of the Queen's palace, and was suitably impressed, but Trades Unions were far from her mind as she began her first week at Somerville Hall. *They* weren't fighting for women, were they? What difference would Unions make to the lives of girls and women? None. It was enough that she had got where she wanted to be for the time being. She knew she was selfish, but was intent upon enjoying her studies. One of her friends from Manchester, Ethel Warburton, was at another hall for ladies in Oxford and so she did not feel completely cut off from her earlier experiences. The young women in Oxford did not talk about beaux or dances; most of them were very serious and Louisa felt she was regarded as rather a frivolous person. It was amazing how differently you were judged in different company.

One morning, five weeks after her arrival, when she had been trying to keep her fire going in the small grate with the coals allowed her and also to decide what further pictures she could put up in her little room, she decided to walk over to 'House,' as it was called and look through the pile of letters for her year on a neat tray. She did not expect anyone had written to her, although Mama now owed a letter, but then saw her name in the neat upright script of – could it be? Yes, it was! Aunt Sarah must have told Miles of her whereabouts. She clutched the letter to her bosom, feeling absurdly happy and took it back to read in private.

> Dear Louisa,
> I was back in Paris, having decided to come back to Europe after all. Father wanted me to enter the Sorbonne. He said two years in Boston were quite enough for me to make up my mind what work I was to do. My mother would have preferred me to stay – but I can always pick up my courses there next year. The great news is that I was invited by Aunt Sarah to stay over with her at Lake Hall Park – which, as you know, is only twenty or thirty miles from your nunnery!

Louisa stopped and looked once more at the stamp. Of course, it was English. He was here in England! She sat down. She *did* want to see him again. As yet she had met no kindred spirit and certainly no young men. They were, she had discovered, absolutely taboo to young women in Oxford.

> So here I am! Aunt Sarah says she will write to your Principal and ask her – that is if you would like it – if you might come for a weekend here. I don't know whether they let you out in term time. I am told it is very strict for young ladies. If not, well, I could come over and you might ask permission to talk in the parlour or whatever they call it with your friend,
> Miles Demaine

He signed his name simply, without a flourish. The ball was, it appeared, in her court until Aunt Sarah wrote to Miss Lefevre.

Oh dear, it was *not* allowed to leave the Hall in term time. It would have to be when term was over. But she did want to see Miles.

Strangely enough Paris seemed even more remote from Oxford

than it had from Langthwaite. Oxford was a town built by and conceived for men. Ladies were tolerated, but only just and if they were to do well in their studies they must stay put for eight weeks. The term was short enough. She took up her pen and began a reply immediately.

> Dear Miles,
> How absolutely ripping to hear from you. Please tell Aunt Sarah that they do not allow us out even on visits to relatives unless there is something really urgent like a funeral. I'm sure we could talk 'in the parlour' tho' it would be rather glacial. But if you are still to be there at the beginning of December I could come and stay for a weekend. Perhaps Aunt Sarah will have left by then? I am very busy with lectures. We are allowed to attend if we sit at the back. Even busier with private tutorials which take a lot of study if one is not to disgrace oneself. But I will not miss seeing you and the Boissiers, I assure you. Let me know.
>
> Your friend
> Louisa Mary Mortimer

She signed *her* name with a flourish and sat looking over the autumn garden and wondering what they would think of each other when they did meet.

The reply was not long in coming. Miles was engaged in private study himself at Lake Hall Park and Sarah was staying indefinitely. She had been very tired after looking after her Aunt Susan and now her own mother was ill and needed her. So Olivia and she were to stay till Christmas at least. Sarah wrote too in her big sprawling hand.

> Dearest Lu,
> What a shame they will not let you out. So you must write and ask your Papa if you may come for a few days before Christmas. I shall be here till the New Year and you will be company for Olivia, who misses Paris.

All that remained now was to ask her parents' permission to stay with Sarah, which surely they would not withhold. Louisa settled down to write to her mother and mentioned casually at the end that Aunt Sarah had invited her at the end of term to Lake Hall Park to see Miles and Olivia, who were staying there. She continued her letter with a long description of her day and the work she was doing at the time. It happened to be the rather tedious translating of an Anglo-Norman epic. She liked the story but found the labour involved in translating it painfully line by line without the crib exceedingly tiresome. There was so much to read in the modern French authors and here she was studying some ancient courtly romance.

'I really don't see what you have against it,' said Ned.

Louisa's letter lay between them on the breakfast table. Emma was looking from one parent to the other, round-eyed.

Marianne tried to remain calm. 'Oh, of course I suppose there's nothing in it. But I wouldn't have thought that now was a good time to be gadding about seeing young men. After all, she has her work to do.'

'I expect she will not neglect that,' replied Ned, who had great faith in his daughter.

'It's just – well, Kate has such an odd life, so rich, and we don't know her son at all,' Marianne continued rather lamely.

'We could invite him here, I suppose?' suggested Ned. This was not what Marianne wanted. That they had met at all was the problem. Even so, she trusted Louisa more than she would have trusted herself at the same age. Louisa was more confident, had a higher opinion of herself. It was Miles Demaine she did not trust. Brought up by Guy, who did not have Ned's attitudes to girls and women, there was no knowing what might happen.

'Sarah will be there – it's only for a weekend,' said Ned. 'She's probably met lots of young ladies' brothers in Oxford.'

'I don't know about that,' said Marianne. They were both silent for a moment.

'He sounds intelligent,' said Ned tentatively after the silence had prolonged itself.

Marianne was thinking, I shall have to tell Sarah one day if they are to go on meeting.

'Yes, I'm sure he is,' she said, 'but he's been brought up as a Catholic, as I said before – and the French don't have the same attitude to women as Englishmen have.' She felt she had rehearsed this conversation many times. 'Louisa mustn't think she is suited to a roving life on the Continent with a family with more money than sense,' she said. 'Of course I don't want her to miss enjoying herself or meeting people. She would be out now if she hadn't refused all that. We must see about introducing her to some nice boys up here.'

Ned looked sceptical, but said, 'Well, I have no objection to her going in December to Sarah's – Sarah's girl is there, too. Perhaps young Miles has his eye on *her*.'

'Yes, I hadn't thought of that,' said Marianne, a little cheered.

But she went in to Cook with the arrangements for luncheon with a heavy heart.

Ned went off to Manchester. He trusted his Lulu. Of course, he didn't want her getting involved with Kate's son, but knew you could not dictate to her. One must just wait and see. Marianne seemed a little too much against the lad himself. Why should he be at all like his mother? He had disliked Kate, but his instincts for fair play inclined him to give Miles the benefit of the doubt.

Marianne had deliberately not said anything about Ned's obvious past disapproval of Kate, reserving that argument for later.

Louisa had the distinct impression during a lecture on the *Chanson de Roland* that some presence was looking down on her from the gallery of the lecture hall. When she turned her head, though, she saw no one she knew. On coming out of the building where the lectures – for the most part appallingly delivered and appallingly boring – were held, she thought once more that out of the corner of her eye she had seen some face staring at her, a face which then dissolved into a crowd of visitors. Who could it have been? Her thoughts were so much on Miles and the projected weekend at Lake Hall Park, that she wondered if he could have come sneaking up to Oxford. But Miles would make himself known to her. It was only eighteen months since she had seen him, but he could not have changed all that much in the interval.

Her parents had still not replied to her letter, which was unusual for her mother, who always answered letters by return of post or the next day. Mama had not approved of the religion in which Miles had, according to her, been brought up, and neither did she approve of young men on the whole. She was odd about them, Louisa thought. Not that she had yet given her mother any cause for concern over any unsuitable friendships. And Miles was not unsuitable. After all he was the son of one of her mother's great friends and well known to her other great friend. Kate Demaine was an odd choice for Mama to be friends with, she reflected, but then they did not see each other now. Chance threw all manner of people into one's way in youth and the fact that her Mama and Aunt Sarah and Kate Demaine had come out together, seemed still to be of some significance to them. Would her own new friends in Oxford fade away as she grew older? Was it just chance that had thrown them together? Louisa made friends easily and sometimes had cause to regret her too easily given confidences. There were many pleasant and some not so pleasant young women in Somerville Hall and she supposed one could cultivate those who appealed and ignore the rest after a few weeks.

Marianne's letter came the next week when autumn seemed to have suddenly changed into winter and the damp Oxford climate had begun to seep into their skins. Louisa, who liked warmth, was crouching and shivering rather dramatically over her small fire. On a cushion by her side lay a novel which was not on her course of study but which she would far rather read than anything that was, and further away on another cushion a pile of unsorted notes which were, she hoped, about to provide the material for her next essay. They were certainly all kept busy and she managed to prepare her work satisfactorily on the whole but she needed the novel-reading as an occasional escape from too present reality. She reread the letter, using it now as a firescreen. They had nothing against her spending a day or two at Lake Hall Park, though they hoped she would be sensible and take some work with her. Her trunk must be dispatched home for the Christmas vacation and she must remember to take suitable clothes down to Sarah's in case there was a house party. Mama hoped Louisa would not neglect Sarah and her daughter. There was something her mother was not quite saying and Louisa pondered it. Was she frightened that her daughter might be about to start an *affaire du coeur*? That was stupid. She looked forward so much to talking to Miles. She wasn't lonely, but she could not imagine ever not looking forward to talking to Miles. Was Mama frightened that a young man would deflect her from her studies, or fearful that she would not be chaperoned? That was all nonsense. Sarah was there.

Louisa put the letter down and reflected (as young women are wont to do) upon her own nature and whether her mother would ever be right to be a little anxious. It must be because she had never shown any interest, as far as they knew, in any young man before. They would surely not imagine that Miles Demaine was a moonstruck young calf who was about to steal Louisa away from herself? Nonsense again. She would tell them everything that befell her at Lake Hall Park, for all was above board. Miles was going to be her *friend*. Letters were no substitute for personal contact. That was the reason she wanted to see him. To get to know him better. What could be wrong with that? She might not even like him so much on further acquaintance!

Term came to an end before the new students had realised that it was not going to last for ever. If the other eight terms passed so quickly they would be grown up before they realised it, thought Louisa. She had sent her large trunk on home and said goodbye to her new friends, who were off immediately with parents and guardians to their own homes up and down England. She would be home, too, by next week, but for the present she was going to Sarah's to see, once more, that fabulous house and its parkland. She was excited. She changed stations some miles to the north of Oxford, managed to get herself on to another local train and was quite enjoying the journey when it stopped and she got down at Bancote to find a pony and trap waiting in the station yard. The groom stood by, but the young man who advanced, hand outstretched, was Miles Demaine. He seemed even taller, even more handsome, as he took charge of things and smiled and laughed and chattered as the trap ambled slowly down to a crossroads. There they followed a lane for a mile or two that led directly to the park. The path then was winding and bumpy. She had never been to this part of the property before and looked from side to side as they passed the end of the lake and a little wood. Then the house came into sight with the bare trees guarding it. After thanking Miles for meeting her she had decided not to try any subject of conversation too quickly but to sit back and enjoy the rest of the ride. He talked enough for two. But she could not resist remarking that the place looked a little neglected. It was out before she intended.

'Alas,' he said. 'I guess Aunt Sarah's family have a hard time keeping it all going, you know – she asked me to apologise for not sending the big carriage.' He sat opposite Louisa, beaming away, and Louisa smiled, saying, 'Then I shan't need to dress up? Mama thought there might be a house party.'

'Oh, no, it's just "family" – and I have to go home to Paris soon. Oh, Louisa, I am so glad you could come. There is such a lot to say, isn't there? My, you look a regular English girl here! Do you like Oxford, then?'

'Oh, I do – ' She began to answer his questions and they were deep in talk before she realised they were about to approach the front entrance and the carriage turned to go in at the old stable courtyard. He helped her down with her little boxes and said, 'Follow me.' Louisa followed, seeing everywhere evidence of work needing to be done on the fabric of the building. But it was romantic, very beautiful, and she was going to spend a splendid weekend. She smiled back at him as he gave a slight bow and disappeared, leaving her to be greeted by Sarah and her daughter. The former appeared delighted to see her.

'We are so glad that you could come. Miles insisted upon fetching you. I do hope you didn't mind the trap – the carriage is being repaired. Mama is upstairs – not too well, I'm afraid – and my husband is staying in Paris till Christmas, but Olivia will show you the park and then we can all eat together in the little dining-room.

A maid appeared to whisk away her outdoor clothes and Louisa stepped into a great hall, obviously part of the original house, where a bright fire was burning and a little table with sherry waiting for the traveller.

'Unless you want to rest?' said Sarah. 'It's not far, I suppose, Oxford, but it's an awkward journey. Never mind, come and get warm. Have you met my brother Ralph?' A tall, ugly man, with slicked-down hair unwound himself from one of the chairs by the fire and lazily extended a hand. He said nothing, but continued to peruse a volume of sporting prints. Olivia came up again and said little, but Louisa could feel herself being surreptitiously regarded by her two sharp eyes. She was probably pulling her clothes to pieces, Louisa thought ruefully.

It was all such a contrast with her own, more homely, home. Aunt Sarah was doing her best to warm up the place, but that was not easy. How had she felt when she came here the first time ten years before? It had seemed different then and she did not remember feeling cold then either.

'Is the ballroom still there? And the corridor where we went skating? And the bedrooms up at the top? I *do* remember, you know.'

'I'll show you round,' said Olivia, quite kindly.

'I'm afraid we don't keep it all as well as we ought,' added Sarah, and Louisa remembered Miles's allusion to financial difficulties.

'Now when you've had your drink, why not come up and say hello to my mother?' suggested Sarah, consulting the watch that hung on a little ribbon on her chest. Louisa was a charming girl, she was thinking, and had grown up since her stay in Paris. But any girl of Ned's would be nice to know. Dear Ned! She thrust the thought away from her. It was childish and disloyal.

Louisa wondered where Miles had got to. He was clearly being polite, not monopolising Sarah's guest, allowing her to become acclimatised before claiming the long talks he had promised her on the way there. Thoughtful boy, she said to herself, and then, No, not a boy, a young man. I think he really likes me, was her next thought, as she followed Olivia up the broad staircase that swept up under a great arch supported by pillars going up from the central hall. On either side of the stairs facing the wrought-iron balustrade of the staircase were hung small watercolours. But once up on the next floor it was darker, and a long corridor ran along behind a brown-painted wall with doors at the far end. She looked back and saw another staircase leading upwards. The place was huge. She didn't recall this part of the house from her previous visit. No wonder it took a lot of money to run. The 'skating rink' must be at the other end on the ground floor. Olivia knocked briskly at her grandmother's door and a maid opened it to them with a finger at her lips. 'She's asleep,' she whispered.

'No, I'm not,' muttered a voice, and the maid looked resigned and made way for the two of them. Lady Gibbs was in a high-backed chair by a fire with her feet up on a footstool, and her grey hair, profuse for such an old lady, spread on the cushions behind. She did not look at all well.

'Brought Miss Mortimer to see you, Grandmama,' said Olivia in her abrupt way.

'Miss Mortimer . . . oh, my dear – Ned's girl – how nice!' said Lady Gibbs. 'They did say you might be coming.'

Louisa bowed and said, 'It is most kind of you to invite me.' This lady was not at all like her half-sister Susan in Paris. Louisa had noticed that old Madame Boissier in spite of her physical weaknesses was in fact a very tough old lady and that she much enjoyed having her son and her family running round after her. Even Uncle René had appeared a little bit frightened of her though he had also looked annoyed. He never argued with her though, and never defended his wife or his children when the old lady was critical, which she often was. But *this* old lady was different, much nicer.

'I'm sorry – there are not many horses,' Lady Gibbs went on, under the impression that Louisa had come to ride. That was what girls usually did. 'And you must thank my daughter, my dear – she has all the organisation of the house. I'm afraid I'm no good for anything much. Do come nearer and let me look at you. You've got your mother's colouring. Do you remember when you came here before? It does seem a long time. And there's Miles, too – off in the library, I expect. Olivia, my dear, would you help me arrange my shawl?'

Olivia moved up and deftly tidied the old lady's shawl and set her cap aright.

How old was Lady Gibbs, Louisa wondered. She must be about seventy. Her own grandmother Amberson was not quite so old, but grumbled a good deal. How sad it was. She could not imagine being old herself. She must make a special effort to be pleasant and kind.

Lady Gibbs turned her face to Louisa again for a moment and Louisa smiled. She is charming, thought Penelope. Reminds me of someone, not Ned – can't think who – but a definite look of Marianne Amberson too.

The maid stoked up the fire. A wind could be heard whistling further up the chimney and Olivia excused herself. Louisa sat on telling the old woman about Oxford and how she had been so happy in Paris too, and how kind Sarah was and how hospitable, and then she remembered that her mother and father had wished especially to be remembered to Lady Gibbs and dutifully mentioned this. After a silence, when Louisa thought Lady Gibbs was once more asleep, the old lady started up again. 'The American girl's son – a good boy, I think – d'ye know him?' She had forgotten that she had just mentioned Miles.

'Yes, Sarah introduced us in Paris,' replied Louisa, with a slight start.

'You'll find him in the library – your mother liked the library,' she repeated.

'No Rollerama this time,' smiled Louisa.

'What? – oh! Fancy your remembering – I do declare that was my idea!' She chuckled. 'When you get old, my dear, you forget some things, but others you always remember.'

'Yes, I always remember Aunt Sarah's Christmas tree,' sighed Louisa. 'So very beautiful – I love this house.'

'Do you? It's too much for my son. Sarah tries her best – we've only two horses now – I shan't see it any better in my lifetime.' She was silent. Louisa took her leave, then, leaving the old woman to rest.

'Look for him in the library,' mumbled Penelope, half asleep.

She had intended at first to follow the old lady's injunction, but then changed her mind, so met Miles again only when the other members of the family gathered for dinner. She had wanted to wash, and dress her hair. Sarah's elderly maid had been waiting to see if she could be of any use, so Louisa made use of her. She only wondered at herself afterwards. One did not usually change and do one's hair for a talk in a library. She rationalised this by deciding that Miles had work of his own to do and would not want her butting in. At dinner it was different again. Sarah and her daughter said little. Lady Gibbs did not come down and Ralph, Sarah's morose brother, spoke not a word, though he was at the head of the table. He did not seem to mind that Louisa and Miles chattered away and excused himself early: 'Have to see a man about a horse.' Louisa could not believe it!

The three young people gathered in the small sitting-room and Sarah joined them with her eternal tapestry work, but did not take much part in the conversation.

'Miles has been looking forward to your visit,' said Olivia. 'He has tired of trying to talk to me about furniture and wants something more intellectual.'

Then she took up one of her books of samples and began to draw from it, leaving Louisa to conduct the 'interesting' conversation. But Miles himself seemed strangely subdued at first.

'Shall we go for a walk round the park?' suggested Louisa. 'Or is it too cold?' Surely they would not object that she walked in a garden with an almost cousin. Or did she need a chaperon to look at the lake and pass comments on the architecture?

'If you like,' said Miles, looking at her closely. 'But why not play chess with me? We must prove our intellectual pedigree.'

Louisa was happy to oblige. She was a tolerable chess player, but as it turned out the game was pretty even. She had his bishop, he her rook. Then they looked up from the board and she felt as though they had never been separated. Their conversation flowed, but Miles seemed more interested in her recent exploits, less in her reading, though he gave that a quiet attention. Slowly Louisa realised he was trying to find out something. Could it be that he wished to know if she were 'attached'? She saw him glancing once or twice at her left hand where she wore a pearl ring her father had given her on the third finger. He remarked upon it eventually. 'That is a pretty ring – was it an heirloom?'

'It was my Papa's mother's ring,' she was pleased to tell him, and she saw what she imagined was a look of relief on his face. Could it be that he really did like her? She must put him to the test. She began to tease him about the attractions of Italy and the comparison between Italian beauty and English. Did he not prefer something wilder and darker in feature of both landscape and human than the landscape greens and the pinks and yellows of England? Were not the olive faces and black hair of Italians and the browns of the Tuscany landscape and the vast palazzas with their splendour and classical elegance grander than the pretty rambling gardens and fields of her own country?

'You are arguing against yourself,' he said. 'England started the formal landscapes in imitation of the French, and Italy *is* wild in parts, but so surely is your own country of Westmorland – smaller, but just as wild.'

'I was wondering if you found the English rather what the French call *fade*,' she asked slyly.

'No! England is small, but beautiful and so are her young ladies,' he said in an exaggeratedly Continental manner, with a bow. Then he laughed. She laughed with him.

They returned to chess for a moment, but she saw him looking at her with a quizzical expression. She was no good at flirtation. It was fun, but palled after a time.

'Were you at a lecture in October in Oxford about courtly love?' she asked innocently.

'No, why? Did you think you saw me?'

'N-o-oo – I thought I saw someone I knew – looking down from the gallery and later in the square, but when I looked round there was no one.'

'We did not go to Oxford in October,' said Miles. 'I thought it better to leave you to your studies. That would be about the time Monsieur Boissier came to stay and he may have visited the university. He gets restless here and often goes out for the day to Oxford or London.'

Louisa looked at the chessboard and realised that Miles's next move would lose her the game. Provided he made the move. Perhaps he had not seen it. She concentrated on willing him to ignore it so that she might have another chance. But did she like the sort of young man who would let a woman win? No. *Had* he seen it? His hand came out, hesitated, and then he looked up at her.

'Go on,' she said. 'You can win the game now.'

'Did you arrange for me to win it?' he asked.

'Oh, no – I hate losing, but you had better defeat me, hadn't you?'

Then he made a move which even more economically defeated her, a move she had not foreseen, and then he looked up at her, and burst out laughing.

'That was not the move I was thinking of. That was better,' she said composedly. 'Ah, well, you will improve my game.'

'Thank you,' said Miles. 'I enjoyed that.'

'Let's take candles and go to look at the corridor where we roller-skated,' Louisa suggested, feeling suddenly restless. 'Come on, Miles, you need exercise.'

Olivia put away her drawing book and rose. 'I shall say goodnight, Mama,' she said.

'Goodnight, darling. Now, Louisa, you know your room. They'll bring a bath for you in the morning.'

'We thought we would look at the Rollerama,' said Miles, taking up a candelabra.

Louisa said goodnight to Sarah and followed him out of the room across the great hall and to a door opposite. Miles opened it, said, 'Here we are – remember?' He held his candles high, for there was no gas lighting and she saw the wide linoleum-covered passage. Yet it was smaller than she remembered it.

Louisa looked out of the window. 'It was snowing before,' she said. 'We made a snowman.'

'Tomorrow we shall go for a long walk,' said Miles. 'English girls like to exercise, and so do I.'

Louisa laughed. 'Goodnight,' she said. 'I'm sleepy. See you in the morning.'

He remained standing with his candle, looking at her as she wove her way back and into the hall. He had so longed to see her again and she was not a disappointment. But he was a well-balanced young man and knew he must spend some time getting to know her. She was so attractive – there must be other men who thought so too.

Louisa for her part felt happy and calm. Miles was even more pleasant than she had expected. They would have many talks before she went away and perhaps they could write to each other regularly. She would leave it to him. He seemed to want her to be his friend.

17

It had occurred to Miles Demaine that what he was beginning to feel for Louisa Mortimer might be what people called love. It had taken him some time to realise that love was this strange mixture of pleasure and pain. When he had met her in Paris and when he had written letters to her, there had been only the feeling that she was interesting. Attractive, of course, but he had admired many attractive girls before. Now his feelings were strange. Even if he discounted the rather disconcerting physical evidence he found in himself when she was near him, there was still something else. He did not quite know what to do. Louisa did not seem aware of any disturbance to his equilibrium as they tramped round the garden on the next day of her visit and then sat comfortably talking in the little sitting-room. She was not the sort of girl he had imagined desiring. He had had daydreams in the past of a sophisticated older woman who would relieve him of his virginity or of a fresh-faced young Italian peasant girl with whom he might find pleasure in the olive groves, but he had never put a face to his dreams. Louisa was a real person, not just a face, a girl with a mind, and just as solid as he was.

As for Louisa herself she was used to being overcome by sudden realisations, sudden desires and equally sudden enthusiasms. She found as they walked together that the young man who had at first appeared pleasant and agreeable to talk to and to think about was quite suddenly a creature of mystery and myth. How could he have changed overnight into a person whom she wanted to know as well as she knew herself? Louisa was, like her mother Marianne, a romantic, but in Louisa's case she had grown up almost unaware of the fact. There had never been too much opposition to her wants and hopes; her parents had loved her. She knew she was different from her brother and sister, but had put it down privately and a little shamefacedly to a superiority of intellect. Now it seemed that her feelings were more interesting and more upsetting than they had ever been in the past. She was even rather cross with herself, as well as amazed. She wanted to find out what he had been doing all his life. Details of his two years in America, which before would merely have interested her mildly, now assumed importance. She saw his handsomeness as something almost alien. She wanted to remain Louisa, not become Louisa-thinking-about-Miles. She had no idea that he was thinking about her. What was there to interest anyone in her apart from her ideas and her tendency to assert her opinions rather strongly? There was no mystery about her, unless all people were mysterious in ways they had never realised. Miles made her feel both more alive and more aware, but at the same time she determined not to show him anything of her new feelings which might lower her in his eyes. She did not expect him to reciprocate them.

The two therefore spent a rather unsatisfactory morning in the stables and the grounds and were relieved when Olivia joined them in their walking and exploring. Olivia restored ordinary life and enabled them to say things without ulterior motive. How long this restraint would have gone on was unclear. Olivia went in to undergo a serious hair-brushing from her mother's maid and they were left once more together, standing by a fountain that was empty of water in a cold wind that whipped Louisa's cheeks to ruby red and emphasised Miles's clear-cut features as he leaned against the wind, covertly looking at his companion. He felt he had traversed aeons in a morning. 'You look rather sad,' she remarked without thinking, perhaps unconsciously wanting some expression of his real self in his reply.

'I suppose I may be,' he said and looked down to the gravel at their feet.

'You are cold,' she said kindly.

'Louisa – '

'Yes?'

'Do you think we might see each other when Christmas is over – you don't go back to Oxford straight after Christmas, do you?'

Her heart jumped, but she controlled herself. 'I have to go home,' she sighed. 'I *want* to go home, of course. I wish I could ask you to stay with us, but – '

'I'd love to, but you're not sure your parents –?'

'My parents are rather strict,' she lied. 'They want me to get on with my work. It's quite an achievement to have been able to go and study away from home, you know – for a girl, I mean – and I don't want to upset them . . .'

He wanted to go on seeing her. That was something anyway. In truth she did not care a whit about upsetting her parents. How could she upset them? She worked hard, did more or less as they wanted and anything else was an extra that they had no business to enquire about. In any case, it was Mama's friends she was staying with. Papa's too. What could be more commonplace?

He was silent for a time. 'I wouldn't want to upset them. Of course you must study. I just mean that – well, I'd like to see your home and your parents, you know. I've always wanted to see the north of England. And my mother always spoke of yours as an interesting woman.'

'Did she? I suppose Mama *is* quite interesting – though I wouldn't have thought your Mama had really much in common with her. Your mother is really very sophisticated, isn't she? Mine has had a very different sort of life. The countryside, you know, the provinces, but she doesn't mind it – she likes a quiet life, I think. I know my Papa does.'

'Tell me about your Papa,' he said, ignoring what she had said about Kate, though a little frown puckered his forehead between his eyebrows.

'Oh, he is a dear – works too hard, at least my mother is always saying so. He's very English, very tolerant. Quite radical, I suppose, in a way. I've always liked him better than Mama.' She paused to see what effect this truly indiscreet remark might have upon him. But Miles was either used to direct remarks or chose to ignore it.

'In a way I think Papa is a little like Aunt Sarah, you know, kind and – well – *honourable*.' She blushed. 'I do like my Mama – of course I do – but she is more complicated than my father. What about yours?'

For a moment they were just two young people comparing notes about what they knew best – their families. Louisa had not known you could talk to a young man like this. But Miles was not any young man.

'It's different in Europe. Children don't know their parents so well – they are just adults who are sometimes there, sometimes not. I believe England is more like the States. People complain about American children.'

'I always forget you're American,' she said.

'Well, half – '

'Do you like being two things at once – always gadding off? Isn't it all rather upsetting?'

'I'm used to it, I guess. It's nice talking to you, Louisa – I do enjoy it. I know I can be rather boring myself – you must stop me if I bore you.'

'Oh, no, you don't,' was out before Louisa could reflect that she might gain an advantage over him if she pretended to care nothing for his company. But that would be absurd. He could be quite a challenge. He was not the sort of man who would always agree with people if he thought them stupid. At least the ice was broken and for the rest of the weekend they did not need Olivia's presence to make them feel less awkward.

Miles observed Louisa when she had no idea he was looking at her. She was not vain and not self-conscious, or at least no more so than most educated girls. *She* was worrying more about her mind than her looks. Wouldn't Miles think her rather ill-educated if he knew of the gaps in her knowledge? He seemed to know such a lot about so many things. But he didn't show off his knowledge, just obliged with answers to her questions and seemed to enjoy enlightening her. He really was rather a dear – well more than that, but she stopped herself thinking too much in that direction. The hours that they passed in each other's company sped like lightning. Miles made no attempt to flirt or to take any liberties. He was always perfectly proper, perfectly pleasant, and Louisa had no idea of the turmoil going on in his discreet bosom. For Miles had learnt early to conceal his impulses. With a mother like Kate who could be quixotic, he had, as a small child, learnt to depend upon himself. That was why he felt somewhat at a loss to find how much he cared for someone else's opinion, the someone else being Louisa Mortimer.

All too soon Louisa had to return home for Christmas. It was the first time in her life that she had not looked forward to seeing her parents and brother and sister and dear old Langthwaite again. Miles had acted very circumspectly but she was sure on the day of her departure that he felt more for her than friendship and interest. Her own feelings had been growing apace. Miles had become not just himself, the tall young man who talked with her and teased her and joked with her and with whom she found she liked to argue a little, but also an idea, a presence. Almost she felt he was more real when she could sit quietly in her bedroom and think about him. That was absurd! Then she would go down to Sarah's family and find him reading or talking to Sarah or see him at a distance walking in the grounds, and the reality of his presence almost seemed to obtrude upon her growing feelings for him. All she knew was that she would miss him when she left Lake Hall Park and that she had never felt like this about anyone else. If only he would say something, even something within the permitted range of statements a young man might say to a girl, but something she could think over later and measure her own thoughts against. He did. On her last morning when she went down to the library to put back the books she had borrowed (for she had managed to do some of the reading that was required of her in the vacation), he came in and stood by the window as she was bending down to return a volume of Béroul, her mind full of Tristan and the love potion he had drunk which had led to his passion for Yseut. She stood up and then joined him at the window embrasure.

'It's going to be colder – do you think it will snow for Christmas like it did before?' she asked in a small voice.

He turned, and the look he gave her was so full of longing, almost of trepidation too that she added swiftly, 'You do remember?' She looked down, not able to bear the weight of feeling that seemed to pass from his eyes to hers. Yet she might be mistaken. Why could she not say herself, I think I could love you, Miles Demaine – I think I may love you – yes I love you.

Then he said, 'Would you mind if I wrote to you? Would you like that? I shall be returning to Paris soon myself and I shall miss you, Louisa.'

'Oh yes, please write,' she answered breathlessly. '*I* shall miss *you*, Miles, truly I shall. I wish we were not both going away.' For a moment they held each other's gaze before he took her hand and softly held it. Nothing else. Perhaps he was not sure of her.

But it was a delicate avowal, she was certain. She said, 'Let us promise to write to each other, and perhaps I shall be able to come to Paris again next year.'

'Oh, do you think you could? That would be wonderful!' His

face broke into a smile and then they both smiled. Some sort of compact had been made in that moment.

He let her hand drop when he heard Olivia's voice. Olivia came into the room shortly after as they still stood looking out over the gardens and the path that wound over to the lake, saying no more, but both a little surer that something had happened between them. Olivia soon shattered the precious moment by asking Louisa to get her bags ready for the old housemaid to take down to the trap.

Louisa did not want Miles to see her off. She wanted her last image to be of him standing at the window. But Sarah decided they would all accompany her to the little station whence Louisa would take the train to Bancote and another to Birmingham where Hugh, who was journeying home from Rugby, would meet her.

So Louisa went home for Christmas, both wildly happy and terribly sad and Miles began the first of his letters to her, not being able to bear her absence and trying thereby to keep the thread of connection alive and growing.

With a newly acquired discretion, which was foreign to Louisa's nature, she played down her impressions of Miles to Marianne and Ned and spoke instead of Sarah and Olivia and Lady Gibbs and the library with great fervour. Somehow Mama did not approve of Miles and now was not the time to have an argument. Miles could arrange some meeting with her parents in the not too distant future, she felt, and then they would see what a wonderful young man he was!

The Christmas festivities passed in the usual way, sanctified by childhood memories. Louisa spent most of her vacation afterwards bickering amiably with Hugh and working away diligently, but her private diary extended itself to a length hitherto not achieved. He had said he would write to Oxford not to Westmorland at her insistence, though he had wanted to write straight away, so she returned to Oxford with the double hope of a letter and of more time to herself to think about him when the constraints of family life were removed.

She found when she returned that absence had made her heart grow certainly fonder and she read his first letter with many sighs and much excitement. Sarah was still at Lake Hall Park and wrote in reply to Louisa's bread-and-butter letter with an artless pleasure. That Louisa had so enjoyed her little stay was indeed pleasing and Lady Gibbs had remarked on her very favourably. They had had a letter from Miles Demaine also, who was still not sure what he was to do in Paris. She had the idea he really rather liked England, though the Gloucestershire countryside must be a dull place for a young man. Even so, he would probably return to the States if his father allowed it. She had written to Marianne also and she did hope the Mortimers were well, for Marianne had said that Ned seemed tired. Had there been a return of his lung troubles? 'I expect your Papa works too hard as all men seem to do nowadays. He will be glad when your brother can join him in the business, I imagine.'

Louisa had a little compunction when she read this. She had not noticed that her father appeared tired, though she knew her mother worried about him. She had not really noticed much at home, truth to tell, her thoughts being too full of Miles for there to be much room left for anyone or anything else.

She went back to Miles's letter and reread it once more. She had almost committed it to memory when the second letter arrived. It was not exactly a love-letter, but there was enough in it to allow her to imagine what he might write one day or what he was thinking underneath his words. He quoted from several poets, she was pleased to see. Mr Browning seemed to be a favourite and this led her to go to the college library to search out all the Browning it held. She would have liked to reply with quotations from 'Tristran', but this seemed a little strong, so she contented herself with searching in Browning herself for an adequate hint to her feelings. She hoped he read her own hidden message. She did not quite dare quote from 'Two in the Compagna' which said exactly what she wished to say, so alluded to its being at present her favourite poem and hoped he might quote it back. How she would love to walk in the Campagna with Miles. How she would love to hear him speak Italian. Instead she quoted 'Home Thoughts from Abroad', feeling that he might feel the same about England. In her commonplace book she copied out the 'two hearts beating each to each', of Browning's 'Meeting at Night', and sat for long hours dreaming over it and over other poems about lost loves and partings. How stupid she was, she thought, even while copying them. For Miles was not yet her lover and she had not lost him, only just begun to find him. But a certain melancholy seemed appropriate when she thought of Miles far away, though she was surprised to find this melancholy in herself. Was it part of love? Did she truly love him then? She pulled herself together to concentrate upon her work and managed to put him out of her head for a quarter of an hour. Work should enable her to see things in perspective, but on the other hand it also appeared to bore her when all she wanted was to continue her acquaintance with Miles and allow things to take whatever was to be their natural course.

The Hilary term was well into the fifth week of a cold and rainy winter and Louisa, who had written a letter to Miles and posted it, was sitting one morning in an armchair by the bright fire in the House common room, toasting her toes preparatory to accompanying her tutor's group of young ladies to a lecture on Aristotle, (though why Aristotle, and what significance he had for her work she could not yet imagine). Miles's last letter had been more forthcoming. He never wrote at great length, but simply and pungently described his thoughts and observations. Quite apart from his feelings and hers, he was a good writer – amusing and earnest at once. Louisa had just decided to put on her cape and galoshes to go to find the other young ladies and was in the entrance hall, when a maid came running down from the Principal's rooms, which lay directly above. Louisa watched her advance towards the pigeonholes which had recently replaced the pile of letters awaiting young ladies on the special table, to put a yellow envelope into the 'M' aperture. Then another maid followed her, more sedately and, catching sight of Louisa, said, 'Miss Mortimer? Take that telegram out, Patty – she's to come to see Miss Lefevre.'

Louisa started. 'Me? What is it?'

The older maid replied, 'Come with me, Miss. It's lucky you were here. Patty took the telegram, but there's another for the Principal.'

Louisa's hand flew to her throat. All she could think of was Miles. He'd had an accident! Her heart began to beat painfully, almost choking her.

She climbed the stairs, following the second maid, who thrust the telegram in her hand saying, 'Miss Lefevre said she wants to see you before you open it,' and was ushered into the Principal's sitting-room, which she had only seen once before. Miss Lefevre was sitting at her desk, but rose when the maid announced Louisa.

'Miss Mortimer, do come in. I'm sorry Patty took the telegram for you. You have it? My dear – it's your father, your mother telegraphed to me at the same time – '

'My father?'

'Do sit down, Miss Mortimer.' Miss Lefevre gestured her to a place by another brightly glowing fire. 'Your mother says, – ' she pointed to an open telegram on the small table by the fire next to a book of Greek verse, which Louisa saw she had been reading. 'Your mother sent me this this morning. Shall I read it to you?'

'Yes, please,' said Louisa weakly, all thoughts of Miles suddenly gone.

The Principal put on her gold-rimmed spectacles. ' "Please advise Louisa Mortimer her father had a stroke yesterday and allow her to return home immediately. Marianne Mortimer." Now you may open yours,' she said kindly.

Louisa's trembling hands tore open the yellow envelope. 'Papa ill – better come home. Take train arriving Kendal midnight Wednesday. Mama.'

'I'm sorry,' said Miss Lefevre. 'You must go home, my dear. Don't worry. Perkins will arrange for a cab to the station – you may have permission to go. I'm sorry,' she said again.

Her mother would not ask her to come home unless it was serious, thought Louisa. Poor Papa. She rose in agitation. 'There is a train to Birmingham at one o'clock. I have written the connection down for you. Let us know of your arrival. And stay as long as necessary,' added the Principal.

'Thank you, Miss Lefevre.'

'Leave your things – just pack a simple bag. Have you money for porters?'

Louisa went down to the hall and ran to her rooms, passing the students who were to attend the lecture with Miss Shaw and leaving a message to say what had happened with the most sensible of them, Edith Hope.

Then she stuffed her sponge bag and a change of clothes and Miles's letters and her diary and her purse with a few sovereigns, which Papa, dear Papa, had given her at Christmas and delivered herself over to Perkins, the head porter, who had a cab waiting for her. A journey of eleven hours would not be fun. Then she thought, Fun? And Papa is ill. A stroke? What does that mean exactly? She was too agitated to eat, though she had found a packet of sandwiches thoughtfully provided by the Principal's bursar waiting for her in the porter's office. All she could think of was that Papa was ill when all she had been thinking of was Miles. And her mother had told Sarah that he was tired again. Oh, Papa! Please God, don't let anything happen to Papa. All through the nightmare journey north she half prayed, half dozed, but made her connection in time so that when, a little late, the train, the fifth she had taken, steamed slowly into Kendal at midnight, she was half hoping it had all been a mistake. But one glance at her brother Hugh's face told her there was no mistake.

Marianne had stayed by Ned's side from the moment he had staggered home on the Monday evening saying he did not feel well and collapsed on the floor of the bedroom just as she was about to help him to bed. The doctor had been and gone three times since then. Hugh, sent for on the Tuesday evening, had arrived back from school white-faced, and Emma had taken over the servants and allowed her mother to stay with Ned. Only early on the Wednesday had Marianne decided to telegraph to Louisa, having felt Hugh must be told first. Ned had drifted in and out of consciousness, occasionally muttering a name or a word. 'Louisa,' he had said and frowned and then relapsed once more, his breathing stertorous. Doctor Thorp had sent for a surgeon from Manchester and they had conferred together and summoned Marianne to hear that Thorp had been right. It was a stroke. Ned might recover, though one side was paralysed. All would depend on whether another stroke supervened. All they could do was wait. Marianne had refused to leave her husband and was sitting by the bed holding his good hand when Louisa arrived. She went straight up to her, frightened and agitated. Marianne barely acknowledged her presence. Her face was drawn.

'I can sit with him, Mama,' Louisa whispered. 'You must get some sleep.'

Marianne finally looked up. 'No, no, you must be hungry – get Polly to fetch you some soup, or Emma.'

'I'm so sorry, Mother,' whispered Louisa. She stood for a moment staring at the grey face on the pillow and a wave of love for her father swept over her. She did not cry and neither did Marianne.

'We can only wait,' said Marianne. 'Go, now. Thank God you are home.'

'Promise to fetch me if – he gets worse,' said Louisa, her hand on the doorknob. She felt she ought to stay, but felt equally strongly that she could not bear it.

'Yes, darling.' Marianne turned once more to Ned and wiped his face with a cologne handkerchief. It was useless, but one must do something. All through the rest of that night, while Louisa slept off her own exhaustion, she sat by Ned's side. If only she could tell him. But she must not. If he died she would have to go on bearing it alone. 'Oh, Ned, Ned,' she murmured, 'I do love you. Forgive me.' He opened his eyes at one moment towards morning and seemed to stare at her. The only words he said then were, 'Love,' and then 'Louisa,' and 'Marianne.' The last with a sigh of satisfaction. The children were all with him throughout the next afternoon and Doctor Thorp was once more in the house. Marianne was willing him to wake and be once again her Ned, when he opened his eyes once more and looked at his wife with a sort of recognition, before giving a convulsive spasm. Then his face stiffened and he died.

In the midst of the horror and the sorrow and her own tumultuous feelings, Marianne remembered that Sarah was still in England and asked Louisa to telegraph her. The funeral was to be on the following Monday. Hugh and Emma were stunned, hopeless, unable to believe that their father had gone. Marianne went to bed, slept a little, but arose the next day calm and drained. It was Alexander Carmichael who came to help out with the arrangements to bury his old friend.

When Sarah heard the news she wrote immediately to René and explained she would be further delayed in England. Her heart was heavy, not only because Ned had died and Marianne would be grief-stricken, but utterly abandoned to grief herself. She had always loved Ned Mortimer – had loved his daughter partly because of him. She set off immediately for Langthwaite, leaving the capable Olivia to look after Lady Gibbs.

When she arrived she found Marianne still calm on the surface, Louisa less so. The house was shrouded in mist and the winds bitter. Everything seemed to be waiting for the funeral. Once her husband was buried, Marianne thought she could try to come to terms with his death. Until then she felt only horror mingled with her grief at the suddenness of his passing, reproaching herself inwardly for not noticing any signs of his impending end. 'No, Mrs Mortimer,' Thorp had said, and she went over his words obsessively, 'this was nothing to do with his earlier illness. Cerebral thrombosis can strike at any time – even a man like your husband in the prime of life. He didn't eat or drink to excess, led a busy life. There was nothing anyone could have done, believe me.'

Alexander Carmichael remained steadily with her, seeing solicitors, sorting out bills for settlement later. Marianne accepted his help humbly. Indeed, her apparent calm was numbness. Louisa found herself wishing her mother would give way to her grief and then she and her brother and sister would have felt less inhibited from weeping themselves. Louisa cried in the privacy of her room every evening and there Sarah found her on the Friday.

'Is Uncle René coming to the funeral?' she asked her.

'No dear, but he told Kate apparently and they are sending Miles. I've just heard.'

'Miles, coming here!' Louisa got up in her agitation and sudden joy.

'René says Kate insisted he come to represent his parents. Kate was off to spend the rest of the winter in the south with Guy. My husband says Miles will stay over at Lake Hall Park on Saturday and then come on the Sunday train in time for the funeral.'

'Have you told Mother yet?'

'No, I've only just got the letter.'

'Tell her, then. Please tell her. She might not want – *I'm* glad he's coming. I can't believe all this has happened. Only last week, only a few days ago, I was in Oxford never imagining that Papa was even ill – I heard from Miles, you know. We write to each other. Miles always wanted to see this part of the country,' she said disjointedly.

She seemed overwrought and Sarah left her. Emma had been helping her mother to be fitted for mourning clothes and Hugh had gone over to Kendal with Mr Carmichael to order the coffin. The house was deathly quiet when Hugh and Alexander returned with the undertaker's men. Sarah found Marianne alone in her little morning-room, pacing up and down. In the past few days she had grown thinner.

'Marianne, René can't come. He sends his apologies,' Sarah began. 'But Kate is sending her son – he'll be arriving at Lake Hall tomorrow. I didn't ask René to ask Kate, she offered herself. Louisa seemed to think you might mind. I hope it's all right?'

'Kate's son is coming here?' Marianne stopped her pacing and looked stunned.

'Yes, dear. He is a nice young man. Louisa will have told you all about him, I expect.'

'Why should Kate want to send him? Couldn't she come herself? Not that I expected her to – ' said Marianne. She sat down and buried her face in her hands.

'If you don't wish him to come,' began Sarah. 'I'm sorry – I told René, of course, immediately I received your letter. I wish *he* would come, but you have all your neighbours and friends around you – and your children – I will try to help you. I expect Kate thought someone should represent them. Or Miles may have offered to come himself.'

'Sarah, if he is coming, very well. But afterwards, after the – funeral, there is something I must tell you or I shall go mad. I expect Miles is wanting to see my daughter and any excuse will do!'

'Marianne, dear – you are overwrought. You will feel better when you can mourn quietly alone. I didn't send for Olivia, I thought she had better stay with Mama – but if there is anything else?'

Marianne gathered herself together and said, 'The firm is sending a group of employees and undermanagers. Alexander is seeing to the gathering afterward. I can't – I can't,' she began again. 'I can't think of anything until – my dear Ned, my dear, dear Ned.' She burst out crying at last, almost glad to be able to give way to grief instead of dwelling on the awful unmentionable truth which had begun to obsess her cruelly now that Ned was dead. And now Guy Demaine's son was coming to torture her.

Sarah left her after covering her with a cashmere shawl and seeing she was comfortable on the *chaise-longue*. She went out wondering if perhaps Ned was in debt or bankrupt. Miles would be no trouble, might even help Louisa. She had noticed they seemed to like each other. Though a funeral was hardly the place or the time, she was hopeful that the presence of Miles would enable Louisa to get through the funeral without breaking down. It was uncharacteristic of Marianne to have spoken of the young man so dismissively.

Louisa was torn by the conflict between her joy at Miles's unexpected arrival – and a heavy, miserable foggy state of mind. She still could not believe he was dead. Her dear, kind Papa. She had seen him laid out. The first dead person she had ever seen and he did not look at all like Papa. His face was grey and his eyes closed. It was horrible. The horror of death was more to her than her grief. If only she could go back a week in time and everything would be as before. But nothing ever would be the same. What did studies matter and ambitions when people died? Had she been kind to Papa over Christmas? She searched her conscience. She had had one or two nice talks with him before she returned to Oxford.

The weekend dragged, but everything was ready for the funeral; the flowers ordered, the food prepared for afterwards by the servants. The state of the family finances, explained in a letter from the solicitor, was waiting for after the funeral, lying on Marianne's little rosewood desk. Louisa was waiting on the Sunday night for Miles to arrive from the station. Other friends of her parents who lived too far away to drive over on the Monday were staying in the little tourist hotel down by the lakeside. She must do the honours and everything must be seemly.

Marianne was, in fact, the first to see the tall young man striding along, dressed in sober black. She watched him from the window. If only there were a respite from her guilt. Never would there be the opportunity now to confess to her Ned who his daughter really was. Thank God Guy Demaine had not taken it into his head to accompany his son. Why should Kate bother to send Miles unless perhaps she knew that Louisa was this Miles's sister? She watched him come up the drive from the carriage, looking round with an appraising air. She must be polite to him. Only after the funeral, when he had gone back and Louisa too had returned to Oxford might she unburden herself to someone. To Sarah. Till then she gritted her teeth and tried to think of her other children and of Ned who would soon be in his grave.

18

The darkly clad figures clustered round the open grave in the churchyard. Emma was crying on her mother's arm. Hugh was standing stiffly on Marianne's other side, a puzzled frown on his face. Other friends and neighbours and the group from Manchester stood a little way off. Louisa also stood a little apart from the rest, next to Sarah who was openly weeping. The men were hatless, and rain began as the well-known words were intoned by the parson. He had liked Ned, although he had disagreed with him over matters of church and state. Louisa swallowed as the first clods of earth were thrown into the grave. Miles Demaine stood respectfully to one side, moved more than he had expected. He had said little to Louisa, only looked at her gravely and with frank sympathy. He is helping me to bear it, she thought. In spite of her sorrow for her father, Louisa couldn't help thinking that her mother would see how nice Miles was and agree that he might – that is if Miles wanted to – Here Louisa stopped her thoughts once more and tried to concentrate on the service. She simply could not believe that her father was in this box that was being lowered into a hole in the ground, that now he was alone for ever, suffocated by the clods of damp earth that thudded over the coffin. 'In sure and certain hope of resurrection . . .' She looked up again and saw Miles, his head now bowed. Then he too looked up and, although he could hardly smile at her on such an occasion, she felt the warm current of feeling between herself and him like scalding tears, that would break the ice-floe in her heart.

She and her mother were alone in the parlour. The other guests had gone except for Sarah, who was walking in the garden. Hugh and Emma were in the kitchen with the servants, who were also grief-stricken and worried that they would soon be losing their jobs. Until the mistress spoke, no one liked to mention the possibility.

'Louisa, I must ask you – for your own sake – not to encourage Miles Demaine,' Marianne was saying.

Louisa stood stiffly and uncomprehendingly. She was off back to Oxford in her mourning clothes on the morrow. Mama had shaken hands with Miles and been polite but distant to him, and then Miles had gone, and now she was alone with a mother whom she simply did not understand.

'But, Mama, nothing has happened. It's just that we get on well together – I know you would like him. Aunt Sarah does. It was so kind of him to come. Papa would have liked him, too. I haven't done anything wrong. Why are you so against him?'

Marianne felt the unseemliness of the occasion, but was determined to warn her daughter off the young man.

'I'm sure he *is* very pleasant, but things can get out of hand. You know your own nature – I couldn't help seeing the way you were looking at him.'

'No, no, Mama! I had thoughts only for Papa. You are wrong. But before – at Aunt Sarah's – we did talk, yes, and we do like each other. He has gone away – I shan't see him again for ages, I expect. Please, Mama, why are you angry?'

Marianne tried another tack. 'No, no, I'm not *angry*, Louisa, I just don't want you getting involved at your age, that's all. It's because I love you.' If only I could tell you why, she thought. But now it would seem the worst possible slur on a dead man's reputation. She sat in her funeral clothes, feeling as cold as ice. Alone now. 'We'll say no more about it, Lulu – just remember not to encourage him,' she said finally and Louisa went up to her room feeling doubly bereaved. She had been so sure that Mama would like Miles. She had said goodbye to him and he had squeezed her hand and said he would write soon again and she must remember he was thinking about her. And he was so sorry and grieved on her account. Her father must have been a very good man. It was such a pity. But he would like to help her to recover – perhaps in the summer she would come to Paris? Then he was off and she had felt the comfort of his presence for an hour or two before the interview with Mama. She gave up trying to understand and packed a suitcase for Oxford with a very heavy heart. Mama had not even wanted her to travel with Miles. So she was returning alone.

As Louisa neared the end of her return journey, Sarah and Marianne were sitting in Marianne's bedroom. A fire had been lighted and the wind was howling through the pine trees that had been planted around the house fifty years before. Hugh had returned to school and Emma gone on a visit to the Carmichaels. Apart from the servants the house was now empty. Marianne had read the solicitor's letter and Ned's will. She had begun to work out whether she might afford to go on living at Langthwaite.

Now she had Sarah to herself and had begun on the confession which should have been made over twenty years before. She hesitated, not only because of the momentous nature of that confession, but because it would seal off for ever her happy married life. But what else could she do? Her daughter's possible feelings for Kate's son must be nipped in the bud at whatever cost. Still, she could not bear to hurt Louisa. Was it a coward's way out to ask her oldest friend to tell the young Miles himself?

Sarah was in Ned's big chair where Marianne had asked her to sit. The curtains were drawn and Marianne had walked to the other end of the room.

'I said I had something to tell you, Sarah. I should have told you, or someone, years ago. It won't make any difference to the past, but I can't bear it any longer. Now Ned is gone I feel I am suddenly back twenty years.' Yes, it was true. She did feel that. Sarah made a move as if to say, That is natural, do not distress yourself, but Marianne said 'No – let me speak. You may not want to stay here with me after I have spoken.' She moved over to the fireplace and stood, one hand on the chimneypiece, her head down, looking unseeingly into the fire. 'You may wonder, Sarah, why I didn't want Louisa to travel back with Miles Demaine – in spite of Louisa's meeting him in Paris and staying with you at Christmas when he was there. I'm not being an old-fashioned mother, Sarah. I would give anything in the world, except my daughter herself, to change what happened and what I have to confess to you. Ned never knew. No one knows but me and now you are to know and then I can mourn my husband as I want to and must. Oh, Sarah!' She looked up, her face distorted with pain and shame. Then she said, 'Ned was not Louisa's father, Sarah.'

Sarah, who had expected nothing like this, started up in order to take Marianne's hand. Marianne looked away.

'No, no, hear me out. You remember my silly girlish passion – though it didn't seem like that then – for Miles's father, for Guy?

When I married Ned I was pregnant with Guy's child. Ned's asking me to marry him saved my life, I think. Louisa is Guy's, you see. They are brother and sister and so they cannot be allowed even to begin to like each other.'

Sarah was stunned and looked at Marianne with such a pitiful expression of concern that Marianne came up and sat down beside her.

'I could never tell Ned. He loved Louisa. And Louisa – it would ruin her life to know too late. All I can do now is to ask you to tell Miles and to swear him to secrecy. Please, ask him never to see Louisa again. I know she could love him and he is, as Louisa keeps saying, such a nice young man. Now what do you think of me?' She turned her face away. After a time she felt Sarah's hand on hers. 'Don't ask the details – spare me that – but will you, could you, do this for me? Tell him what I have told you. But no one else must know. Not René or Kate or Guy.'

'Did Ned never guess?' asked Sarah at last in a husky voice.

'No, never. I don't think it crossed his mind, you know. And as for Guy – it was all over before it had begun. It was my fault.'

'And you have borne this secret all these years alone! Oh, Marianne, my dear! You should have told me before and I would have seen that they never met. But now – I am an old goose, but I think they do like each other – if perhaps they could know each other as brother and sister?'

'No, it is not possible. I know my daughter . . . the laws of the church and of humanity itself . . . could anything be more terrible? I wanted to tell Ned. Often and often I have braced myself to tell him. I don't know how I managed all these years. By "forgetting", I suppose, except when we were at Lake Hall Park when they were children and I saw then how it might turn out.'

'It was my fault for introducing them,' cried Sarah.

'No, of course not. But you see Louisa is a little like me – rash – '

'Miles is not at all like Guy!' said Sarah. 'He is a fine boy. I don't know how Kate produced such a good person. Isn't there any way it could be avoided? If we just asked them – I could talk to Louisa about it?'

'No, I mean, I'm sure she would listen, but would not understand. And I cannot tell her. I know I ought to. Perhaps she would not even blame me, but to take away the memories of her father, the man she has always thought was her father, I could not do it. I wish I lived in a different society, Sarah. Why should I have paid in misery? I have been miserable for so long – and now Ned is dead and I can't bear it. It began to alter the whole idea of our marriage the moment he died not knowing the truth. I sat by him when he was unconscious and I tried again and again to summon up the courage to tell him. I knew it would be a relief to me, but would send him to his grave in misery.' She began to cry.

'I will do what you have asked,' said Sarah, standing up. 'I will tell Miles the truth and he will know what to do. Matters have not gone very far between them, I'm sure. He could go to the States – I think he wanted to anyway, but Guy wanted him in France. He will be of age in a year or two and his parents will – or his mother will – settle money on him and he can go away. He is too honourable to exploit his knowledge.'

'That is what I thought. He is the sort of boy I'd have chosen myself for my daughter. I could see that even yesterday. And the one man she cannot have.'

'What will Louisa do, though? These things hit women harder than men, I think.'

'She will finish her studies and learn to deal with a broken heart as I did,' said Marianne. 'But it is because I love her I want the best for her. Ned agreed when I said that I wouldn't want our daughter to marry Kate's son – a Catholic, and so young – but in the end he would not have opposed it, I think. Perhaps it is a mercy he has gone. To punish me too. Oh, Ned! Ned!' She burst now into noisy weeping. At last the floodgates of her grief opened and Sarah saw her to her bed. She herself lay long and wakeful that night. She must not fail her friend. But, oh, she wished it was not she who would have to deal the blow.

It was so unlike Marianne. Of Kate she could have believed it, but not of Marianne. She went over that summer of their first season, putting together things she had noticed at the time, but forgotten. To marry Ned, Ned whom she herself had loved far more than she could ever have loved René Boissier. She had made a mistake too and was paying for it in her own misery. But in comparison with Marianne it was nothing, only the cross so many women had to bear. And Ned had not asked her to marry him. He would have done though, if Marianne had not accepted him. If Guy had not seduced Marianne, she herself would now be the sorrowing widow, not Marianne. Dear Ned. And dear Louisa, whom she had thought so like her mother. That was why. Marianne recognised that her daughter too might give way to passion and provoke a lifetime's unhappiness, she supposed.

It would be done. As soon as she returned to France, which would be now within a week or two, Lady Gibbs being a little better and René waiting for her to take up the reins of her household. Of the three of them she thought only Kate had really got what she wanted. Kate and Guy. They must never know.

Kate, however, had had her suspicions about Louisa Mortimer from the very first time she had met her at Sarah's in Paris. Alone of them all, she had seen the likeness between Louisa and her husband. They were confirmed by Guy one day of early spring in Monte Carlo. He had been a little drunk and had had a good day with Kate's money at the gambling tables. Not that Kate ever gave him much. Just enough to keep him happy without ruining her. Guy was not averse to spending her money – his 'pocket money' he called it – and he was under the impression that by gambling he was adding to the family income. He knew that the system of *séparation de biens* as practised in France, prevented him from laying his hands on the greater part of her capital, but the money allowed them both for living expenses, for travel and clothes and furnishing and servants and houses was rather a tidy sum, and he had never pushed his luck by asking for more. So long as he had a small amount for his own use. He had long ago stopped earning anything himself and Kate did not mind – or did not appear to.

That March afternoon when they were both sunning themselves in the glass-covered hotel rooms that looked out over the bay, he had returned with a full purse and an expression of satisfaction. The champagne he had drunk was still enlivening his imagination and he could be a witty and amusing companion. The conversation got round to poor Ned Mortimer's demise. Miles had written to them from Paris and had described the funeral. 'I believe they were awfully pleased I went, though Mrs Mortimer seemed hardly aware of anything, so shocked was she by her husband's death. Little Louisa was pale and pretty and I met her sister and brother. Hugh is a pleasant chap, but has not much to say for himself and Emma is a Hausfrau already. They live simply – not many servants.

I was sorry never to have met Louisa's father for he seems to have been a fine man from all accounts.'

Guy had read the letter that lay on the bureau in Kate's bedroom. They kept separate rooms by mutual agreement.

'Marianne Amberson was a passionate little thing,' he said and smiled complacently. Kate, who was drinking Russian tea, paused in the act of lifting the spoon to her mouth. 'Of course, she wasn't one of those prim English Misses,' he went on reminiscently.

'What do you mean?' she asked.

He did not notice the narrowing of her eyes, for he was usually unaware of his wife's thoughts and she never cared to enlighten him. But today he was in an expansive mood. 'Only that she once succumbed to my charms,' he said.

She raised her eyebrows. 'Did you think of marrying her, then?' she asked in a bored way.

'*Mon dieu, non*! It was only the once – an afternoon at Lake Hall Park – you were out riding with Mortimer, I do remember that. It was a bit awkward, I hadn't expected it, you see.'

'Poor Guy, did she seduce you?'

He laughed. 'Oh, no, she may have thought she did – certainly she made it clear she wanted me.'

'That was not sensible.'

'No harm done, probably did her good. She was odd, though.'

'Odd to want you, you mean? Of course you can be very seductive, *chéri*.'

Kate knew Marianne to have been a passionate and impetuous girl and could only imagine what agony she must have gone through. She ought to have had the sense not to allow Guy Demaine liberties which Kate had certainly not allowed him before marriage.

'Perhaps she *did* seduce me!' Guy went on.

'Ah, well, what is life without a gamble or two,' she said. 'I know she liked you, of course. What if she had had a fortune and I had been poor?' Would he have married Marianne?' She thought not.

'Girls with money have more sense,' replied Guy.

Kate looked at him as he ordered a *sirop de cassis* for himself. They got on well, she and this charming but weak man. Marianne would have been miserable with him. But why should Guy suddenly feel the urge to confess?

'Why didn't you tell me before?' she asked.

'I'd forgotten about it to tell you the truth. Marianne was not the only girl who was mad about me.' He laughed. 'Of course, I never shared her feelings. I remember old René warning me off her. English girls are funny creatures. I wonder what her daughter's like.'

'Little Louisa? I met her at Sarah's. I told you. Miles seems quite smitten, I guess.' It had not once occurred to her stupid Guy that Louisa Mortimer might be the result of that forgotten episode. She knew now she must have been right. Marianne would be most upset to think that Miles might entertain a *faiblesse* for her daughter. My God she would!

Kate suddenly realised what Marianne must be going through. Of course, there was nothing definite, but she knew her son almost if not quite as well as she knew her husband and had noticed a mention or two of the fair Louisa already. But Guy must never know. That would be putting the cat among the pigeons with a vengeance. Aloud she said, '*I* was not poor.'

'Ah, no – and haven't I been the best husband you could ever have found?' He need not have told her about Marianne. He must feel very sure of her, Kate thought.

'Of course, dear, if you are happy still with our life?' she asked innocently.

'Couldn't be happier.'

'Miles, of course, must not entertain ideas of settling down yet,' Kate went on.

'No, he's a good boy,' laughed Guy. 'Not like his Papa, eh?'

Kate twitched a little, but knew she had him by the short hairs. Guy was a fool. Really, he did not deserve her. But she could do what she liked as long as she was married to him. The gambling, though, must be firmly supervised.

Back in Oxford Louisa was finding it difficult both to get on with her studies and also to feel any great interest in the topics of the day. She had made friends with some of the less solemn of her fellow undergraduates in Somerville Hall, but even Emmeline and Eleanor, two girls studying French like her were shy of offering their sympathy over her father's death, and Louisa was not an easy girl to comfort, being too proud probably for her own good. Three other friends, Ettie, Dora and Grace, were more robust. Ettie was studying Greek and was a clever woman: Dora was a History exhibitioner and little Grace Wilkinson, a clergyman's daughter, was studying Natural Science. The four of them often gathered in each other's rooms for discussions on life and work. Occasionally the talk would creep round to the enigma of Men. And more often to the allied subject of clothes. Louisa professed a disregard for fashion, but always looked well, if a trifle bedraggled. The girls all wore their new tailor-made coats and skirts, their 'costumes', and many were the difficulties of keeping themselves neat and clean and mending trailing linings and dusty hems. Keeping abreast of work and looking after one's personal washing and mending left little time for enjoyment, though Ettie and Grace enjoyed their work in a way that made Louisa envious. They had decided opinions on almost everything, pronounced on the fashion for boaters and fringes with the same interest they gave to the writings of William Morris or their leisure reading of the poetry of Rossetti or Meredith or even of Conan Doyle's *Study in Scarlet*. Louisa found them a distraction from her sorrow over her father.

Nothing had prepared her for so missing Miles and this was getting itself mixed up with her grief for her father. She often sat dreaming and going over her mother's words to her when she should have been writing an essay. Grace Wilkinson would try to cheer her up. She was a forthright girl, interested in the work of the rebel Ibsen. She intended to work in the East End after her studies were over. For Grace had kept her father's social conscience though she had had doubts about his theology and was more aware than most of the lives of the poor. Louisa was ashamed sometimes when she realised how little she knew of the lives of others, in spite of being the child of two parents rather Radical and Progressive in outlook. And she was also ashamed of devoting so much time to thinking about Miles. But Miles would make everything all right if he were there. He would know about Ibsen and the class system and the writings of Morris, she was sure. If she had Miles to talk to she would cheer up.

She had waited with impatience for his first letter after his return to Paris. Seeing him at her father's funeral had made him into a person acknowledged by others. He existed then in the real world. Marianne's words rang in her head though. Something about being led astray by good looks or some absurdity. Why, had she even thought of Miles as being 'good-looking'? What was that 'nature' of hers that her mother said she knew? Marianne had seemed distracted by grief and Louisa put a charitable construction upon her mother's outburst.

When his first letter did arrive, she went to her room and read it in peace. She sat at her table in her mourning clothes, and caught sight of herself in her tiny mirror, the college's concession to vanity. She did look glum, she thought. She began to read. First, condolences upon Ned's death and how sorry he was he had never known him. Then an ambiguous passage about wishing she were there, but how he found pleasure in the idea of waiting to see her again. 'I shall wait for you, dear Louisa', he wrote and the words gave her a thrill. Then in the next paragraph: 'It may be earlier than I had imagined for I have decided to study medicine, after all the decisions and counter decisions of the past two years. I could do this in Boston or London. Mama says she now approves. What about London, eh? Wouldn't that be a good idea? Your devoted Miles.' She sat on, conjuring a vision of Miles in London, Miles saving the lives of the suffering, moving among invalids with his sunshiney presence. Yes, indeed! that *was* a good idea. She took up her pen and replied immediately.

Just about the time that Louisa was writing this letter, Sarah was having the most difficult two hours of her life in Paris with Miles himself. She had asked him to meet her in the Bois de Boulogne, but to say nothing to anyone about it, and he had assumed somehow that there was a message from Louisa. They were sitting outside in a small café but at the back under a canopy where no prying eyes might gaze upon them or ears might overhear. Sarah had not known where to suggest their meeting should be. She had toyed with the idea of a church, but had rejected that as too sacrilegious and unnecessarily melodramatic though, God knows, her news was melodrama enough.

They had sat down and she had ordered two coffees and then two brandies, thinking they might be necessary, at least for herself.

'I have to tell you,' she began and she put her hand on his, 'why you must not go on seeing Louisa Mortimer. At least if your feelings may be what her mother imagines . . .'

At the words he went pale. Not go on seeing her? Was this a sentence of death? 'Not see Louisa – ?' he repeated stupidly.

'Of course, as an acquaintance you might see her, I expect,' Sarah went on doggedly. Then she looked up. 'Believe me, Miles,

I hate this. But her mother begged me to tell you and I promised I would. No one else must ever know – especially not Louisa.'

'What is it, Aunt Sarah?' Miles looked frightened.

'Only – your father . . . Well, before Louisa was born, before her mother was married – your father had relations with her mother – ' Sarah blushed. In spite of her sturdy background it was excruciatingly embarrassing to talk of her friends in this way to a male member of another generation. 'The point is,' she went on, 'Louisa is Guy's daughter and so she is your half-sister.'

He sat looking at her incredulously.

'I'm sorry, Miles. It is true. Marianne never told anyone at all. Certainly Ned never knew. But she had to let you know, because she had the idea that Louisa at least had a little *faiblesse* for you. She has no idea whether this is returned, hopes it is not. It must be nipped in the bud, you see. I am so sorry.' She burst out crying and Miles was appalled.

'Are you *sure*?' he said after a pause as Sarah blew her nose. 'What a rotten thing to ask you to do. Why couldn't Mrs Mortimer tell me herself?'

'Oh, Miles, it isn't possible to speak of these things. I am her oldest friend, you see, and she has been so distressed ever since it happened, always worrying that you would meet. I told her it was inevitable. Of course, there may be nothing at all to worry about. You are both young. Just that it's impossible. And you *must* not, we beg you, must not tell your father. He has no idea and must never have. Nor your Mama.'

Miles took the brandy and swallowed it. 'I do have – feelings for Louisa,' he said slowly. 'How am I to hide them? If I see her I should want to tell her why we could not go on seeing each other. How very cruel. I shall have to go away. Yes, I shall have to go to Boston and study there.'

'You do believe me, then?' asked Sarah earnestly, mopping her eyes.

'Yes, Aunt. It is not the sort of thing anyone would invent, is it? But how damnable for her. Why must she not know herself? She is not a baby. She loved her father.' He pondered this. After a silence, Sarah said, 'One day her mother will tell her, I expect. But she cannot bear to at present, having just lost the husband she so loved, who gave Louisa his name. Ned was a dear, good, kind man,' she stopped, thinking of him.

'Poor Sarah – what a thing to ask you to do,' he said again, recovering his self-possession. But his thoughts were in a whirl.

'Write to her, Miles, as soon as possible. Tell her you are going away. I don't know on what terms you are with her – I hope it has not gone too far. It was my fault your meeting. I never knew anything about it till last week. Poor Marianne! It has quite ruined her life. And she cannot bear to confess to anyone else. You do see?'

'How abominable!' said Miles. He wondered how Papa had treated Mrs Mortimer – was it a long-standing arrangement or a mistake of youthful passion? He could understand it, but not condone it. He sometimes found his charming father very difficult to understand.

'Don't think worse of either of them, Miles – these things happen. I'm sorry I had to do this to you.' She rose and they shook hands and she walked away leaving Miles Demaine to sit alone to bear the first real sorrow of his young life.

To keep away from Louisa, but not tell her why! His mind rebelled against it. What if he saw her 'just as a friend' as one might see a male cousin? This was possible, but there would always be Louisa's own feelings to consider. Yet to reject her out of hand was inhuman. He sat at another cafe in the Boulevard St-Germain and pondered. He knew now without doubt that he loved Louisa Mortimer. And the thought of the way he could love her – and she a sister – shamed him. Yet why? Yesterday she had been the same person. But she was his sister! He could not get over that fact. He had a sister already, though little Kitty had never been much of a companion, being too young and flighty. Louisa was already a sort of sister anyway, an intellectual one. Why should all the foul sex business muck it up? But it would. He knew it would. And it was not foul, but glorious. His feelings swung from north to south and back again and by the time he got up, still irresolute, the coffee chits were piled high on their saucer.

Dear Louisa,

I have decided to train in the States instead of London. No time for explanations. Shall sail very soon. It is impossible, Louisa, that we go on seeing each other for the present. Forgive me. You will not understand, but believe me this is the best way. One day I hope I can explain. In the meantime, Louisa, get experience of life and work and don't forget your friend, but only your *friend*;

Miles

Louisa sat numbly. What did it all mean? She was tempted to read it to Ettie and ask her opinion, but discretion forbade her. Something had happened, but she could not imagine what. If he were to sail very soon there was no time to be lost – she must reply to him by return. Was he sorry he had written as he did before? She got out his last letter from under her pillow and read it again. She did not know whether her emotions were ones of sorrow or anger. But in the night she woke crying and knew it was what people called heartbreak. What was the use of anything if Miles were to go away? She would go to London and look for real work. He said she must get experience of life. Very well. She would. She would flee this Oxford where she felt she no longer belonged, and find some terrible toil and work her heart out. She felt that more than a feeling had been nipped in the bud. A person. Who had made him write like that? Her mother? Surely not. If it were Mama he would have said so. She thought of writing to ask Aunt Sarah, but thought she must not worry her. And Sarah would only tell Mama who would say, 'I told you so.' Perhaps he had fallen in love with someone else. She tried to push this thought away.

Poor Miles had made the letter as mild as possible under the circumstances, but Louisa's reply tore at his heart. If only he could tell her the truth. Although nothing had passed between them which anyone might have regarded as a promise or an engagement, both of them had known that something had been set in motion. He saw it as a little spiritual emanation that had hovered over them, waiting to solidify. What could he write to her now? He sat in the Luxembourg Gardens trying to think how he might alleviate her evident misery. What if he were to say that whatever it was that had separated them, it was as much a blow to himself as to her? Would that help? It was too cruel to tell her he could not love her when he had never yet put in words that he did. Nipped in the bud indeed. In the end he didn't write at all, but busied himself preparing for another year in Boston, this time with a definite end in view. But his heart was heavy. He avoided any talk with either his mother or his father, who had returned from the South of France and were off again, this time to a German spa where Kate knew gambling was no longer permitted. His mother gave him an allowance, not an excessive one, and agreed to pay his first two semesters' fees at Boston Medical School. And, more and more, he began to despise his father. Yet if he were Louisa's father too, how could he hate him?

Easter in Langthwaite was a miserable affair without Ned. Marianne was in a state of collapse, attended by Emma, and Hugh was taciturn and spent most of the day out riding. His own future was being arranged. He was to leave Rugby and join the Manchester firm. Apparently Ned's financial affairs were only just in reasonable order for the time being. Marianne was to have the income from capital, the capital itself being left to her in his will, provided it went eventually to her three children. Things would not be easy. Louisa decided to work this to her own advantage.

One look at Louisa's face was enough for Marianne to know that Sarah had spoken to Miles. Louisa said nothing to her mother about Miles until the last day of her holiday, which she had spent working over her books trying to do as Miles had asked her. On her last evening she went in to her mother who still lay on a sofa before the fire, looking weak and ill.

'Mama, I must speak to you.' Louisa sat down at the other end of the sofa and stared into the fire. It was a late spring and they both wore their shawls against the wind that could be heard roaring round the house and gusting out of the fireplace.

Marianne's heart bounded unpleasantly, but she waited in silence for whatever Louisa might be about to say.

'I don't want to continue in Oxford – I am not going to be a scholar. I want to go to London, to work. There is a settlement run partly by our Hall in the East End of London and I think I could find work there.'

Marianne had not expected this. 'My dear child, you have never done anything to prepare for that sort of charity work. Could you not wait till your three years are over? And where would you live, and on what? I can just manage to keep you on at Somerville.'

'Father has left me a little bequest, as you know,' replied Louisa, still not looking at her mother. 'I could use it and my savings and work for the rest.'

'You are not twenty-one till the summer,' said Marianne. 'And even then I should think you would need permission from your father's executors to use any income from your little legacy.'

But Louisa knew she could not really be bothered to argue. 'I have spoken to Mr Carmichael already,' said Louisa coldly. 'He sees no objection. Indeed he thinks it a sensible plan. Many of the lady students take only a year or sometimes two in Oxford. They do not all finish their degrees.'

'Why should you suddenly decide to leave? I thought, Papa too thought, you were happy – it was what you wanted.'

'What does it matter where I am?' cried Louisa. 'Papa wanted me to be happy. Papa would have liked Miles Demaine too, I am sure! But Miles is going to Boston and I shan't see him again, I don't expect, so I'd rather prepare for work on my own. I must be independent. It's what you have always wanted me to be, isn't it? And anyway you can't stop me, Mama, when I am twenty-one.'

At the sound of Miles's name Marianne had turned away. 'I only want your happiness, darling. I'm sure Miles is a clever young man who has his own furrow to plough.' Louisa said nothing, not trusting herself to go on. 'Have you heard recently from Aunt Sarah?' asked Marianne. She wondered if Miles had already left Paris. Marianne had, in fact, received a terse message from Sarah to say Miles had been spoken to, but had heard nothing of his plans.

'No,' Louisa answered shortly, then: 'I believe Mrs Demaine will only be happy when Miles is in the States,' Louisa went on, for she had decided that Kate was at the bottom of Miles's sudden change of tone and his decision to go away.

'Kate will be very anxious for her son to make a career for himself, I am sure,' said Marianne, hoping desperately that Miles had said nothing to his mother about his interview with Sarah. Sarah had promised that no one else should know the secret and she trusted the young man not to give away such a secret to his mother. Kate might be immeasurably distressed to hear of it herself. If Kate knew, then *she* would suffer as Marianne had done. But, after all, daughters were more vulnerable than sons. A young man with prospects would soon find a girl to love. Louisa should perhaps not be too much opposed in her plans. So long as she never saw him again. But at these thoughts she found she had begun to cry. If only she could tell her daughter. What a relief it would be. At least Louisa would know that the young man had not rejected her for herself, only because of the relationship between them. Louisa did not notice her mother's tears and shortly afterwards crept out of her sitting-room. She had decided to harden her heart against all possibilities of sweet feelings or any plans for the future that were not concerned with earning her living as an independent spinster. But her heart could still not help hoping and she found in her dreams at night that Miles returned just as he had been before and she wept in her sleep with joy.

19

It was the middle of June and the Oxford term had just come to its end. Louisa Mortimer, having 'run away' to London, was sitting on a truckle bed in a tiny room of the Docklands Settlement where she had been accepted for voluntary work later in the year. She was not supposed to be in London till autumn when her Great Uncle Lechmere was to let her a room in his house in Hampstead. At her interview some weeks before in Oxford when she had been quizzed by one of the governors of the Settlement, she had discovered to her horror that the work was unpaid. However would she manage? Now that it had come to leaving Oxford she found she did not really want to give it all up but was too proud to go back on her decision. Neither could she bear to go home immediately term ended. She had not told her mother that she would receive no salary in the autumn, but had decided to find some other work before then. Her friend, Grace Wilkinson, had invited her to stay at the beginning of July with her family in a London suburb, but in the meantime Louisa had arrived hot and dusty in London, with nowhere to go. First she must look for a room before looking for work. She went to the Settlement just in case they might find her something for a week or two but nobody lived in the gaunt East End building except the Warden and her husband. Louisa asked if they might find her a bed for the night, for the Warden knew she was to help them in September and would surely put her up till the morrow. She did not want to start spending any of her sovereigns which were in a little bag tied round her waist. That had been Emmeline's idea. Emmeline and the others had tried to dissuade her from going alone to London before Grace could have her visit, but had not succeeded. They had then gone their various ways to country rectories, the houses of doctors in Birmingham suburbs, Cambridge college lodges or the homes of fathers who were barristers in London.

She had sent a telegram to her mother from Oxford to say she would be going to her friend Grace, whom Mama knew was the daughter of a clergyman, so that would be all right for the present since her Mama did not know exactly when term ended. Now the warden of the Settlement clearly did not know what to do with her; nobody was needed for the present. Louisa had told a fib and said she had arrived a day too early to go on to her friends, who were abroad, and could she please stay just for the night. What she was to do or where she was to go before her visit to Grace she had no idea. It was hard to find suitable work. Young ladies did not usually work. But she wanted to be free, free of everyone. Free from Mama, with her nervous anxiety that had set in since her Papa's death, free of Oxford, and most of all free from the memory of Miles. How *could* he have gone away? She was both pained and angry. Not angry with him or angry that he did not love her, but angry with herself that she seemed to have mistaken his interest in her for a burgeoning love. The Warden had finally taken pity on her.

The room they had let her have for the night was tiny, not more than six feet by six. Her thoughts went jerkily back to the long drive in a Hackney carriage which she had taken to the East End through the hundreds of mean little streets where it was obvious ladies did not linger. The Settlement was near the river and she heard the hooters of ships calling through the tiny barred window which was too high up to see anything out of. It was like a prison cell. How could she bear working in this place in September? She had vowed she would stick it out, however uncongenial. Even the Principal had not tried to dissuade her. She wanted to offer up the job, however unpleasant, to Miles. No more Oxford, she had decided. She looked despairingly round the room once more. The only item of furniture, apart from the iron bedstead, was a stand with an ewer of water. No towel, no looking-glass. She had taken her supper of bread and soup in a large room where women and children came every evening for something to fill their bellies, and afterwards there had been a Bible reading. 'Don't go out again, Miss,' the Warden's assistant had said. 'Mrs Bairstow could not take the responsibility.' It would be only sensible to return home and then go on to Grace's and prepare for Hampstead in September, but she did not feel like being sensible. Even so, this place was very depressing.

She only partly undressed, slapped some water on her face and crept on to the bed. She would decide what to do in the morning. Even if it was only a week's freedom, she was her own mistress. She had run away and would only return when she chose. She had slipped out of respectable society, she thought. This cheered her a little and she fell asleep.

'Mama warned me against beauty,' Louisa had written in her diary that term. 'Did she mean something else?' She was to find out soon enough when she left the Settlement the next morning after a mug of tea from the Warden, for whom she peeled potatoes for that evening's soup, since no money had been asked for. Now she must look around London. But she found her hope and optimism severely tested when she tramped down Tooley Street to London Bridge and Borough High Street where she thought there might be a horse omnibus to Westminster. Several men shouted after her and one leered as she passed, and came towards her as if to touch her face, but she stared stonily ahead and he was quickly gone. How could people live amid all this ugliness, with the stink from the Thames and the dust raised by every passing cart?

Louisa had been protected and cherished and, although she knew such places and some unpleasant men existed, she had never seen them before. The faces of the women who passed her were also shut-in, pinched and hard-eyed. She tried to imagine their lives, but could not. Work at the Settlement would teach her, she thought.

There was an omnibus that came trundling along on the cobblestones after she had waited twenty minutes, and she took a fare to the new Charing Cross Station, which was further away than she had imagined. Here all was bustle. She knew it from leaving there for France two years before. She walked down Whitehall and reached Parliament Square and the Houses of Parliament with their Gothic spires. Uncertain where to go next, she walked in what she imagined must be the direction of St James's Park, but lost her way and found herself in Victoria Street. There was still nowhere to sit down and people looked at her oddly. She ought to have gone into the Abbey, but it was too late, she had passed it. She decided to look for a tea-shop, if such existed. The thoroughfare was crowded with carriages and carts and the smell of horse dung was everywhere. She felt hungry. The tea she had drunk for breakfast was much less than her usual rather substantial early morning meal. She was peering up and down the road trying to look as if she knew it well and was waiting for someone, when a gentleman bumped into her and she looked up startled. Then a voice in her ear said, 'Why, Heavens above, it's little Louisa!'

She looked up into the face of René Boissier, who was staring at her with an expression of surprise on his sharp features. 'Oh, Monsieur Boissier,' she cried. Not 'Uncle', she thought. Sarah's husband was somehow not an uncle figure now. 'Is Aunt Sarah here too in London?' she asked.

'No, no – I'm here alone – what are you doing all by yourself in London then?' He stood there in his checked overcoat with a smart cape and a pale brown bowler on his head, perched a little over his ears.

'At present I am looking for a cup of tea,' she said.

'Then I will accompany you to a small café I know,' he said.

She followed him, a little annoyed that her adventure was to be broken into by an acquaintance, but desperate for a drink.

Over a cup of strong coffee, which René insisted she drink rather than tea, to which he gave a rude French name, she told him a little of what she was doing.

He considered her carefully as she spoke. His gaze was frank and appraising. Then, 'I saw you last winter in Oxford,' he said. 'At a lecture.'

So it *had* been someone she knew staring at her. 'Why didn't you come up and say hello?'

'I didn't feel like it,' he answered after a pause. 'How was Lake Hall Park?' he went on. 'I heard you were there.'

Louisa had a little jump in the region of her heart. 'Lovely,' she said simply.

'I hear young Miles was there, too,' he went on.

'Yes, he's gone to the States again – as I expect you know?' she managed to answer.

'So my wife tells me.' There was another silence as he poured her another cup of coffee. 'So, what are you going to do now?' he enquired. 'If I may say so, you don't look very happy – pretty, but not happy.'

'I shall go to visit my friend Grace next week – I've written to Mama,' she said, as though she had to account to this adult for her movements, which was absurd.

'I have a flat near by,' he said carelessly. 'You could stay there if you wanted.'

She had not known René kept rooms in London.

'So Providence has landed you in my lap,' he mused in his slightly accented English. 'How old are you now, Louisa? Let me see – twenty? Twenty-one?'

'Twenty-one,' she said proudly. It was a shocking thought, but she knew he was attracted to her. With the thought came a thought of Miles, but she pushed it away. 'I am *perfectly* happy,' she added, which was not true, but what business of his was it?

'Were you not told that young ladies must not wander round London alone?' he asked, lighting a cigar.

'Of course I was – but why not?'

'I could put you on a train home,' he said reflectively, not answering her question.

'Oh, no – I don't want to go home! It is so dreary now Papa has died,' she replied.

'Yes, I'm very sorry about that. Your father was a fine English gentleman,' he said, but gave the phrase quotation marks.

Louisa wished desperately that her Papa was with her, that she could go home with him and tell him all about her unhappiness over Miles. But Papa was really gone. Gone for ever. What was Monsieur Boissier offering her? She felt rebellious, but also a little afraid.

'Well, if you have nowhere to go and will not go home, you can certainly stay in my flat,' he said again.

Neither do young ladies stay alone in the flats of married gentlemen, she thought, but wondered if she was being presumptuous.

'You remind me of someone, you know, I can't think who. A little like your mother, of course, when she was your age. Well, are you going to accompany me? I have rooms in an apartment just behind Victoria Street. No one will pursue you there. If you can't decide what to do you may stay overnight – and go home on the train in the morning. Or you could sleep in my dressing-room until your friend can put you up.'

She followed him out of the café. 'It's very kind of you,' Louisa began, and then she thought, perhaps it was not kind, perhaps by wandering round London she had put herself beyond the pale of decent society. She thought he was laughing at her. Why did she not just walk away and try to find a job for a week? But what jobs existed for young ladies? Perhaps there was nothing, and she would have to sleep in the park or part with her precious sovereigns which were to see her through the summer and then the autumn, since there would be no money coming from the Settlement, even if she could bear to work there, about which she was now unsure. She felt frightened and depressed. It did occur to her that she could stay perhaps for the present, as she was to stay later in the autumn, in Hampstead with Great Uncle Lechmere, but she did not want that. He was Family, and this was her Adventure.

'You can at least come and see my apartment,' said René. 'Don't look at me like that – I'm not a big bad wolf, though you may be Red Riding Hood.' Now he *was* laughing at her. She knew he had no great opinion of English girls, for she had heard him say so in Paris. Even though his wife was English, and his Mama also.

'How is your mother?' she asked politely, in order to have time to think, for now they were walking past a large store for the third time and she felt it was becoming ludicrous. And at the same time she was saying to herself, If only Miles were here. Miles and I could profit from this freedom in London. But he is in Boston and will never see me again.

'Oh, let us not speak of mothers and fathers,' said René Boissier lightly. She had no wish to discuss her dead Papa with him and was feeling very guilty about how Marianne would react to this London adventure of hers if she ever got to hear of it, so she did not pursue the subject. 'I could take you to the theatre tonight,' he went on. 'And you could have a meal at Wheelers with me. And you might go to a concert or the Royal Academy exhibition when I have business to do.' He spoke casually, made a little holiday sound attractive. But she had no money for concerts. He would expect to pay for her. Perhaps as a man of her father's generation he felt responsible. Was he wanting to look after her? If she did not accept, might she offend him? What was the good of being unconventional and reading Ibsen if you couldn't even accept a helping hand from your mother's old friend? And Miles, dearest Miles, whom she knew she loved, Miles was gone forever. Why not profit from this unusual meeting and grow up at last? She felt fatalistic. Ever since Papa had died and Miles had written to say they must not meet, she had felt bound to follow some awful path that she could not avoid if she were not to be a coward. How strange though that René Boissier should bump into her like that. She had never connected him with London somehow.

René Boissier was not a complete cad and did not at first push his luck. Ever since Louisa had stayed with his family in Paris he had been obsessed by her face, her young, slim body and her *joie de vivre*. He was rash and he could be ruthless, but he did not intend to hurt her. If she stayed with him no one else would ever know. After all, he was almost in *loco parentis*, was he not? The child was alone in London. But it was the most extraordinary coincidence that she should be here when he had a few days to spare. He noted that underneath her spirit of adventure and rebellion there seemed to be a layer of melancholy and wondered why. Some affair of the heart, he supposed. It usually was with women. He had seen her pretty ankles and wrists in Paris – when she had been oblivious of his regard, and found a growing attraction for her. He had seen her serious little face at that lecture where he had gone hoping to catch sight of her, and had been rewarded. Why, he asked himself, should he be limited to conducting exciting, but rather seedy, affairs with young women who possessed not half the charm of Louisa Mortimer and whose personality interested him not one whit? He ought not to seduce her. Yet she might be willing. English girls were far too free and easy. And he might even do her good. So long as he did not fall in love with her. She would be an amusing companion. Anything else would depend on her, he thought Jesuitically.

All Louisa was thinking as she followed René into Thackeray Mansions was 'Why did Miles give me up? What is wrong with me? He does not love me, perhaps never did. I want experience – if he won't love me, what can I do with myself?' To René, when he had sat her down and put her carpet bag in his dressing-room she reiterated only, 'Mama thinks I am already with Grace. I went to the Settlement, but they would not have me until September.' Then she had to explain the Settlement further and René looked sceptical when she said that she intended to leave Oxford and take up work with the down-and-outs. I wish I could forget Miles, she thought, as she chatted away to René. She would like to have unburdened herself to this middle-aged man who sat looking at her across the room. But she was chary of doing so in case he managed to get some hold upon her, seeing her as vulnerable.

As she talked to René, Louisa looked round the apartment. It was a large 'flat' in one of the new blocks of mansion flats off Victoria Street, built of red brick and rather imposing. She had never been in a house like this before and wanted to ask René if he owned the apartment, but it seemed rather rude. The drawing-room where they sat making conversation was large and aesthetically pleasing. One of its sides was almost half filled by a window-frame with pillars at each end. The curtains of brown velvet were drawn back behind the pillars and she saw a wrought-iron balcony through the glass. It was rather like a French room and she supposed he must have chosen it because of that. Did Aunt Sarah ever come here? She rather thought not. In front of the window on the left was a screen of white-painted wood and next to it a trolley bearing a tray with a syphon, glasses, and bottles. A little marquetry table of the sort that her mother loved stood in front of the screen, on it a crystal vase filled with yellow flowers. She wondered who had brought them and arranged the vase for it was not the sort of thing you expected a man to bother with. Several low-backed chairs, upholstered in yellow velveteen, were parked on mustard-coloured carpet of a very advanced design. René stood with his arm on the chimneypiece, behind which was a large looking-glass, into which she saw him glance a little surreptitiously now and again. Two armchairs drawn up to the unlit grate stood on a rather worn Turkey rug, and an ormolu clock and one or two other small tables more or less completed the furnishings. She sat in one of the chairs, facing a table with a gas lamp, in the form of four female figures holding up the light. There seemed to be a picture on the wall next to the fireplace, but she could not see what its subject was without craning her neck, which might look rude. René was smiling at her faintly. She wished there were a fire burning in the grate and looked towards it.

'Are you cold?' he asked kindly. 'The servant will bring coals if you are. It is unseasonable weather.'

The servant was evidently a fixture that went with the apartment, for she heard a man whistling now across the corridor in what she supposed was the kitchen. She remarked the absence of books, which surely said something about René Boissier's character. She could not imagine any apartment or house of this size without them. Perhaps there was a library though. There had been other rooms they had passed as she was led into the mansions.

'Like it?' he asked, seeing her gaze roving round the room.

'I'm not cold,' she said, answering his first question. Then, 'Yes, it is very spacious – I didn't know you had an apartment in London.'

He burst out laughing at that. 'Why should you? As a matter of fact I come here only once or twice a year when I am tired of Paris and when Lake Hall Park is too uncomfortable.'

Lake Hall Park uncomfortable! It was chilly but not, she thought, uncomfortable. She thought the place beautiful. It was where she and Miles had been together, and was therefore sanctified.

'May I use your bathroom?' she asked a little shyly in order to get out of his way for a moment and do a little thinking.

'All modern conveniences,' he said, getting up, taking her to the door and pointing down the long carpeted corridor. 'When you return I suggest we go out for some lunch.'

When she got back, smelling of his cologne soap, he had poured out champagne for her in a wide-brimmed, shallow glass.

'Drink up,' he said. 'Put some roses in your cheeks.'

She had seen in the looking-glass in the palatial bathroom that her cheeks were a little pale. Tramping round London should have given her some colour, but she had seen a wild-eyed girl whose hair needed a good brush. She had found a woman's hairbrush on a table in the bathroom and used it. She took the proffered glass.

It was a good thing that, when all was said and done, he was still a sort of gentleman, he thought. She looked so eminently desirable as she sat down and sipped the pale gold sparkling liquid. If he had picked her up in the street and she had not been a good girl, he would have conquered her then and there. She looked passionate too. Not that that was always attractive in women. René did not even like his mistresses to be too intense, too liable to hand themselves over to men with bated breath as supreme gifts. He liked to make the running himself. Passion from women could be frightening and was unnecessary. Her mother had been a little like that, he remembered. Not to him of course. Looking at her rather boldly as she sat back, now a little more at her ease – the champagne was doing its work, he thought – she could be French. If she wore her clothes with a little more aplomb and shortened her fringe and had better gloves and had not chosen a scarf that went ill with her coat and skirt.

'You could almost be a French girl,' he said aloud.

'I suppose that is a compliment?' she asked rather cheekily.

'Oh, yes, when I say it anyway. Frenchwomen spend a good deal of time learning how to be chic. On the whole Englishwomen exaggerate themselves when they choose their clothes. One must play down colour while bringing one's best feature to the fore,' he replied complacently.

'What is *my* best feature,' she asked boldly.

'Oh, your eyes, without doubt,' he answered.

'I could never be really smart,' said Louisa after having digested this. 'Aunt Kate is smart, isn't she? I suppose she is a sort of ideal woman to you?'

'Kate Demaine? She married my best friend! Yes she is smart, but I don't like that too opulent look – money cannot buy the only kind of charm that matters. Kate is handsome and chic, I grant you, but she is cold.' He recognised the contradiction in his own thinking but did not elaborate.

Louisa was a little thrilled to find that a grown-up man was talking to her as though she were a worthy partner in this sort of conversation. Not the sort of conversation she had ever had before. 'What is the "only sort that matters"?' she asked, her eyes looking rather solemn and, he thought, enticing, over her glass.

'*You* have it,' he answered carelessly. 'Not all young women do. You have more than your mother had – that is why I said you could be French with a little effort.'

'But the charm is not effortful that is found in young people?'

'No, but one must have it to begin with. Later – only a little later – women need knowledge of how to present themselves to best advantage.'

'Olivia is interested in fashion,' she ventured. 'Your daughter,' she added, as he looked blank.

He said, 'Oh, yes, Olivia knows about style though, poor dear, she has no charm.'

How could he say that, she thought indignantly, about his own daughter? 'I'm afraid I don't take a great deal of interest in fashion,' she said rather fiercely. 'I think I am like my Mama in that.'

He changed the subject. 'Didn't you like Oxford, then?' he asked, filling his own and her glass once more.

'Oh, yes, I did, but I don't think I am cut out to be a scholar. I felt – restless.'

'But why on earth should you wish to immure yourself in a

"Settlement"?' He gave the words a disdainful sound. 'I suppose you were brought up to have a little Radical conscience. Is that it?'

Was it that? Not really. But she could not tell him about Miles and how life seemed now too tame and boring and how she must work to expunge her dreams of him. They began to talk of English society, Louisa having defended the Settlement. 'I shall read you a little sermon,' said René. 'There are three types of workers in England.' He liked pontificating, she thought, and leaned back to listen. 'There are those who aspire to become bourgeois and will do so through saving and practising the English and American virtues of work and temperance. Then there are the craftsmen who want a new society built for hard-working men like them. They will stay in the working class to advance the cause of others – they are the idealists. We have them in France too. Then there is your solid mass – the majority who drink and brawl and idle and whom nothing and nobody will ever change. *They* are looked after in "Settlements".' He sat back.

'You say nothing about women,' she remarked after a pause.

'Women follow their husbands, or menfolk in general,' he replied. 'Don't tell me you are a "Feminist" like your dear mother. It is very unwomanly you know.'

She looked a little insulted. He was amused. Women only needed to be conquered and their Radical notions would fly out of the window. He did not say this, but she guessed the line of his thoughts.

'I wish I were a man, then!' she cried, draining her glass.

'That would be a pity,' he said drily. 'Your charm is feminine.'

'Don't men have charm, then?'

'Some, but of course I could not be a judge of that.'

She thought again of Miles. He was charming, more charming than René. But René was intriguing. Perhaps it was because he was older and more experienced. She felt a little shiver of excitement which she tried to suppress. If women could not be ruined through sitting with men and sipping champagne the world would be a better place. Why, she might well be ruined just through coming here with him. It was absurd. Why should she not have a little experience herself?

'Come, we can continue this conversation at luncheon,' he said. He held her coat for her and slipped it over her shoulders, took down a wide-awake hat from the peg in the hall and they went out again. He took her in a hansom to a large glass-walled restaurant not far away where he ordered oysters to begin with and observed Louisa observing the scene.

Over the chicken à la King that followed, he said, 'You know you would be infinitely worth taking trouble over.'

She smiled, feeling rather spoilt and not able to suppress a little satisfaction. It was rather exciting being with René Boissier.

'I should like to buy you clothes,' he murmured. 'I would soon change your notions.'

This she was not so pleased to hear. But how could a young woman, thirty years younger than a debonair, sophisticated Parisian reply without seeming naive or shallow or stupid? Louisa was not stupid and she knew she was playing with fire. The problem was, did she care? Had she ever known her own nature? Here she saw Miles again in her mind's eye and he remained there even when she was replying to René's bantering tone. It *was* pleasant to be taken out to luncheon, to be talked to and considered and flirted with. Was that not what being young was supposed to mean? For a little moment she wondered whether girls were married off so quickly because they might begin actually to enjoy freedom and flirtation and excitement, and even the perils of passion. Why should she feel it would be wrong to give in to René's seductions, apart from her love for Miles Demaine and her concern for Aunt Sarah? Her innate rebelliousness came to her aid here. It was not conventional ideas of virtue, but a demand for freedom that in the end would make up her mind for her. She felt this even as he was entertaining her and flattering her.

'How long have you lived in Thackeray Mansions?' she asked. By this time they were eating profiteroles and ice-cream with a strange liqueur poured over the top that was like nectar. How much must a meal like this cost? Yet her mother and Sarah and Kate had once had this sort of life during their Season in London. Spending money, or more accurately having it spent on them. She was more provincial than her mother must have been. She would like to ask René about her mother.

He replied to her question about the mansion flats with a slight smile. 'I only rent rooms there – I told you – when Paris gets boring. For some years now, if you really want to know.'

She realised he must take women there, would probably rent

the place for that reason alone and, though not shocked, could not suppress distaste on Sarah's behalf. How strange men were. She was sure her own Papa had never indulged himself in such a way. Men seemed to be as various as women in their interest in the opposite sex. Miles came into her mind again unbidden. What sort of man was he? Did *he* flirt with women? Did *he* have love affairs with other girls? She felt insanely jealous. Her face dropped and René noticed.

'Who, then, is your cavalier? You must surely have one with a face like yours?'

'I have no "cavalier",' stated Louisa.

'There was someone – there must have been or why should you look so sorrowful? Are you in love?' he asked matter-of-factly.

At first she made no answer, then, thinking that an avowal of love for some unknown person might both allow her to be of interest to him and still preserve her from any more dangerous feelings, she said, 'Perhaps, but he doesn't love me!'

'Doesn't love you,' echoed René. 'How can that be?'

'Don't laugh at me.'

'Mais, ma chère – je ne ris pas. C'est sérieux alors?'

'I don't know – ' But as she replied she realised she did know. That Miles Demaine still existed somewhere, that she would not be here being entertained by a middle-aged roué – if that was what he was – if Miles were even remotely within reach.

'Never mind,' said René, who had refused ice-cream. He ordered Stilton for himself and tucked into it with a spoon while Louisa, having declined anything more, waited for coffee.

'So, he doesn't love you, and you are in despair,' said René, conversationally. 'So instead of rushing home to Mama you came to London for consolation. You decided to become a nun, to work like *une petite soeur des pauvres*, to forget him?'

This was, although silly, so near the mark that Louisa looked away. 'It is not fair, is it, *ma chère*. But you need not become a nun – '

'I'm not a nun,' she said. 'I just want to be independent.'

'To be independent one has to have money,' replied René, wiping his mouth with the snowy napkin provided. 'That is the only independence, the only freedom. I'm surprised your mother didn't tell you that.'

'Mama never had much money,' she said, and then stopped.

Had René meant her to ask herself a question about her mother? Money was why girls usually married, unless they were themselves rich. 'Of course, if you are rich you don't need to marry,' she mused aloud.

'Were we talking of marriage? If you wish then. Most girls marry. In England they are said to marry for love.'

'I wouldn't marry for anything else.'

'Then you are less intelligent than I thought.'

They went out together. It was after three o'clock.

'I have to go to my tailors,' said René, depositing her at the entrance to the Mansions. 'You may stay here if you wish.' He rang the bell and the porter appeared. 'Here is the key to my apartment – I shall return in an hour or so. If you wish to go to the theatre this evening, I shall get tickets. I believe your Gilbert and Sullivan are playing at the Ambassadors? Shall we go?'

'I have no dress suitable,' she faltered.

'Then we shall sit in the pit. Meanwhile there are some books in my study. Make yourself at home as my wife's mother always says.' And he was gone.

Back in the drawing-room Louisa felt a slight headache fix itself in her skull. How tiresome. It must have been the Chablis. She yawned and went into one of the rooms through the half-open door. It was a small room with an escritoire and a few shelves. She took a book down to read in the armchair. A queer collection of books in both French and English, with titles she had never heard of. What was she doing here? It was absurd. She went up to the glass in the drawing-room over the fireplace and seemed to see René's mocking face behind her. 'Miles does not love you,' she said aloud, 'Miles has gone,' and felt tears welling up into her eyes. She watched them for a moment and turned away, yawned again and sat down.

When René returned he found her asleep in the chair, the book on the table beside her. He stood for a moment looking at her. How easy she would be to seduce, in spite of her romantic attachment to some unknown youth for whom she was clearly pining. She woke with a start to see his eyes gazing into hers. For some reason the phrase 'drowning in his eyes' occurred to her and she averted her gaze.

'I have the tickets,' he said, straightening up. Then as she sat up and brushed hair out of her eyes, he said, 'You could stay with me whenever I am in London you know, if you wished.' He meant her to be his mistress, she knew.

She stared at him. Was that what unhappy girls did? Abandoned by their prince they threw themselves into the arms of the first personable man, especially if that man was an experienced fellow, apparently at their service. It might even allow her to forget Miles, might allow her to enjoy life in a world respectable girls never contemplated: the *demi-monde*. She rather liked the power she seemed to have over René. But she did not find him attractive – not in the way she did Miles. There was no meeting of souls and without that she found it hard to contemplate a meeting of bodies. How close she was though to surrender. In despair, in boredom – her resources about to fail. Was this how girls got themselves into trouble? She supposed it must be. Yet surely not she, not the bluestocking Louisa Mortimer!

'I am looking forward to the theatre,' she said. 'Thank you. But tomorrow I think I shall go home. I – I can't forget him – I am useless without hope. I shall come back to London and do what I said, though.'

'Tell me about him,' said René conversationally, as though his invitation had never been uttered.

'No, I can't.'

'Is he English?' he persisted.

'No – no, he is not.'

'I might have expected an Englishman to give you up,' stated René. 'But if he is not an *Anglais*, it must be more serious. Of course,' he added, 'you needn't be "ruined" if you stayed here with me. No one would know – unless you told them. If it got out, well then, of course, that would be the end of the *jeune femme comme il faut*.'

She felt his eyes raking her from top to toe as though she were naked and it made her feel dizzy.

'You are too good for a *young* man,' he said. He brought a chair up to hers where she still sat, flushed and a little frightened. 'I want you frightfully – as I think you realise,' he whispered. Truly, he felt as though he'd never wanted a girl so much. He didn't suppose they talked of such things where Louisa was brought up. 'Englishmen don't talk about desire,' he said.

'They may not talk of it, but they feel it,' she replied.

He got up and turned his back on her. Strangely enough, he

thought that he was perhaps not worthy of her! Louisa did not really know anything about desire if she was, as he assumed, *une petite vierge*. That excited him unspeakably. He turned to face her again. 'Of course, you might want an "experience",' he said. 'Young women like you – and your mother – want experience. And I'm too old to give anything up for what you call love. It doesn't exist, believe me. It's a pretty idea. Now we shall go to the theatre and when we come back you may give me your answer. There is a room for you. Dress yourself up like a little milliner going to the music-hall and I shall take you out.'

She got up slowly and found the room as he had said with her case on a stand and water for washing and another hairbrush. She heard him talking to his servant and wondered what sort of a man he really was. Was it true that love did not exist? And what had he said about Mama? Yet he seemed to 'love' her. Her headache was thankfully gone and she applied herself to tidying her face and hair and putting on the only dress she had in her case, all the others having been sent on home. It was a plain, grey affair that she had thought suitable for Grace's family. Not a dress for a milliner, more one for a missionary.

In spite of everything she enjoyed the evening, saw many other young women enjoying themselves too on the arms of their beaux or their fiancés or their husbands, and wondered whether life could turn out right after all. Whether she allowed René Boissier liberties or not she would still be the same Louisa. Would it make any difference? Might it not be a way of forgetting and achieving a twilit status that would enable her to free herself for ever and leave her childhood behind? She *could* do it. Yes, she could. But then she heard a sad song on the stage, sung by a dark young man. The music was nostalgic and tugged at her heart. The worldly crowd who enjoyed it with her was just an ordinary London audience who had their own hopes and fears. She was nothing special. But she had been special to Miles. She knew she had. She was not ready to give up the idea of him, not free to give herself to René Boissier. No. However exciting. No.

When they returned and he lit the gas brackets and they sat for a time together René said, 'I think after all you want to be alone tonight? Am I right?'

She was relieved. 'I'm sorry,' she answered. 'I do like you.' But why should she be sorry?

He made a little moue. 'You must not mention that you saw me in London,' he added after a moment. 'I shall not mention it myself. And you have done nothing for nasty tongues to gossip about.'

'It isn't just because it is "wrong",' she said. 'It's because I want a – certain person – and he doesn't want me. But you have been nice to me.' She felt disloyal to Aunt Sarah saying this.

'Because you won't let me have the opportunity of being anything else,' he said sadly. 'Never mind, Louisa Mortimer. By the way, is the young man by any chance the son of my dear old pal Guy Demaine?'

She went pale and looked at him beseechingly.

'I see I am right. I seem to have the gift for clairvoyance tonight. Off to bed, Louisa – spare a thought sometimes for me. I shall put you on the train home tomorrow morning, never fear. But you might invite me to your wedding one day.'

'You are wrong. There will be no wedding,' she said quietly and went to her room.

René who had made what appeared to him a considerable sacrifice was rather pleased with himself. But the next afternoon on the way back from a visit to his barber's on St James's Street, he met a young lady with whom he had had a pleasantly sordid relationship many years before and was unable to stop himself inviting her round to his love nest in the evening. He would not easily forget Louisa, but she was not for him. It was only several days later that he realised of whom she reminded him.

To Louisa as she swayed and bumped north in the smoky train (the sovereigns had had to be broken into, for René had no intention of paying her fare), the past two days had been a revelation, not of London, but of herself. She had found the East End unpleasant and the West End a place where a completely new Louisa might have budded forth. If it had not been for the thought of Miles she might very well have agreed to develop that hidden part of herself that was never spoken of by respectable girls. She had realised feelings and desires in herself that she knew would one day, in the far distant future, be assuaged and used. In the meantime she transferred the new desires to Miles, who was now a sort of absent companion.

Mother would be surprised to see her, but she would do her duty by her and then return to stay with Grace. The only excitement at home would be to observe curates and teach Sunday School. Somehow, it spite of her sense of failure over Miles and over London, she had scored some sort of victory. There was always the secret knowledge that a man had asked her to become his mistress. Mama should learn none of that from her. She knew René would also remain silent. Poor Aunt Sarah – how awful if she discovered. But where was Miles, what was he doing, thinking? Just to see him for a moment, even happy without her, would be all she dared imagine in future.

20

In Boston, Mass., Miles Demaine was sitting in an oak-panelled library with a book propped up on the desk before him. He turned a page of *Materia Medica* and then sighed and looked out of the window. It was raining. Somehow he always expected sun, not rain, in this land of his mother's birth, falling in the distance over the harbour and plopping dispiritedly on the awnings of the grand houses on the Common where he had walked that morning. He was distressed, uneasy; more than that, profoundly depressed. But he was too young to know that he was depressed, felt vaguely angry with an ashy feeling in the pit of his stomach. For six months he had tried not to think of Louisa Mortimer, without success. She was there when he awoke in the morning and there when he finally went late to bed, having tried to tire himself with study. The odd thing was that he missed her, though he had never missed anyone before. Their acquaintance had been brief but deep. The thought was always drifting about in his head that if Louisa was his sister then they could surely live together as brother and sister. Better than this loneliness. All his newest and most precious feelings seemed to have been put on ice that awful morning in the Bois with Aunt Sarah Boissier. His common sense told him that young men did not live with young women whom the world did not know were their sisters, unless the young ladies were to join a twilit half-world of scandal and intrigue. He could not associate Louisa with any of that, but he still felt there must be some solution. He was not used to confronting problems that had no solutions – his youth objected to their very existence.

The revelation about Louisa and also about his own father had naturally shocked him. He had sat down many times to think it through, and had tried to consider whether his shock was only because he had been entertaining non-brotherly thoughts about her. He had decided that it was not. He had an idea that if a fellow did not know a woman was his sister, his feelings would be appropriate to this ignorance. The trouble was that even now, in spite of his unformulated plans to live with her as a Platonic friend, he did not feel like one.

He snapped his book closed. No use trying to do any more work today. He would go for a ride in the cars and let his thoughts be soothed by the contemplation of other lives seen through lighted windows at night, or half-heard conversations of strangers. Did his mother know the truth? That was what he kept on asking himself. He was not surprised that his father had seduced Louisa's mother, though the details refused to become more than a misty picture of some romantic compulsion, with Marianne refusing to marry Guy. He had never felt very close to his father. Plenty of charm, he supposed, beginning now to see his father as a man and no longer as a protector or as superior in knowledge or prowess. *He* was not like Guy. He would not have treated a respectable girl lightly. He was sure Louisa's mother must have loved her seducer. His imagination left off at this point. He did not ask himself what might be Marianne Mortimer's feelings a generation later. He wanted to write to Louisa to tell her the truth, to explain his dreadful behaviour in leaving for the States without another word. But he had promised Sarah and could not break that promise. Life was bleak. He had begun to think tenderly of Louisa and could not behave as if nothing had happened. And underneath the ice he felt as if his own heart was waiting to beat again.

Marianne had sunk into a tearful and prolonged period of mourning. That was how her behaviour was presented to the world and it was part of the truth. Without Ned she felt rudderless, even angry with him for deserting her. But her grief was underpinned by an even more dreadful guilt, a guilt which altered the whole world for her. Ned's death had touched off ancient terrors. She clung to her children, particularly at first to Louisa, when she returned home for a short time. But Louisa had refused to discuss leaving Oxford for work in London, and never once mentioned Miles. She had returned from her summer term in a strange mood which Marianne, preoccupied as she was with her own agonies, scarcely registered at first as new. At the end of a week in which she had dutifully helped her in every possible way, but as though her parent were a sick child who needed a responsible adult to guide her, she had departed again to her friend Grace in London with whom Marianne had thought she had been to stay earlier when term finished. Louisa was also to visit Uncle Lechmere to discuss her future lodgings. She did discuss this in a cool rather bored way that should have warned her mother that she was having second thoughts about her plans to work in London in the autumn. But Marianne could think of nothing but her own shame and guilt.

In spite of Louisa's being so kind and helpful she was actually glad when she departed for Streatham and she was left with Emma and Hugh, who were less demanding of her spirit. They had no idea that anything other than mourning for their father was responsible for her misery, whereas Louisa had seemed to be sharing some secret of her own when she had, from time to time, glanced at her face in repose. At least *that* danger was past. That was what she had prayed for and what seemed to have happened. Only Sarah and Miles knew, and Sarah could be relied upon to keep her secret. As for Miles, he was an unknown quantity. She hoped he would stay in Boston for good and find the girls of his mother's native land more attractive than her daughter. She dreamed of him once or twice, though, and he seemed familiar, kind. Kate could be proud of him. It was not his fault, or Kate's. It was hers. And her thoughts returned to drain her of hope and energy. Only Alexander Carmichael, who called in now and again with his sister to see how they could help her, gave her a little peace of mind. She could talk about Ned with him without blushing and he accompanied her also to the grave in the little churchyard that stood in the middle of the village and the spire of whose church could be seen reflected in the lake on fine sunny days. She did not want to visit the churchyard alone. The very gravestone was a reproach. She felt she was going mad.

Kate Demaine was usually averse to prying into her children's possessions. Kitty and Miles had been brought up as fairly self-contained people, for it had never suited Kate to claim too much of their attention when she did not wish to give them her own. It was a little unusual then that she should find herself one September day in her son's room in Paris. She had gone in search of a guide-book which Guy insisted he had lent Miles and which Miles had not returned. The room had been tidied up and dusted when Miles returned to Boston to prepare for entrance to Medical School. His

Italian hat and his summer clothes from the year before were still hanging in the large Empire wardrobe and his books filled many shelves and bookcases along with some of the mementos of childhood that he wished to keep – his geological specimens and small plaster figurines. She stood idly looking out of his window that looked out of the inner courtyard over to an old well that still stood under a tree by the ground floor servants' apartments. It was a still, early autumn day and Paris was growling in the distance. Tonight she was to go to a reception on the Boulevard St-Germain where she hoped that the Prince de Vrigne would be present so that she might quiz him over his son. She hoped the son would take a fancy to Kitty next year.

Kate yawned and stood for a moment at the window thinking of Kitty. Then she turned to scan the bookshelves. Guy was very tedious if something was missing. He asserted what authority he had in small matters and was forever admonishing Kitty about the trail of discarded objects she left behind her. *La Princesse de Clèves, Les Fables de Lafontaine, Middlemarch*, Emerson – her son's taste in reading was eclectic. He was a clever boy. She espied the missing yellow-backed guide to the quattrocento stuck behind a volume of Shakespeare on the shelf below, and bent to retrieve it. She took it out and saw a piece of paper stuck in the book as though Miles had thrust it there hurriedly when someone had come into the room. It was in his handwriting and she opened it curiously. Kate was not one to have qualms over reading letters meant for others. It looked as though it was the draft of a letter and she took it to the light, for her fine dark eyes were getting a little long-sighted now that she had reached her fortieth year.

'My dear Louisa' (crossed out) – 'Dear Louisa, Please do not ask me why but we must not meet again for some time at least' (the latter phrase also crossed out and another substituted). 'It is not that I am not fond of you' (that too was crossed out) 'It is only that I think we are too young' (that had been scored out heavily) 'that I have decided my work is to be that of a doctor and the best place to study would be the States.' 'Do not forget me' was then repeated twice and crossed out with a faint wavy line. Under that a bracket and a 'How can I tell her when I *cannot* tell her? What does Sarah want me to tell her? It is impossible. Cruel to be kind?' And then a line of verse from Leopardi's *Primo amore*. She folded the paper and put it back in another book then took it out again, tore it into little pieces, and put it in her pocket to put in the stove.

Kate sat down heavily on the sofa which she had had covered for Miles in dark blue velvet. Her thoughts were quite steady. Sarah clearly knew about Guy and Marianne. Had Marianne put her up to telling Miles about the relationship? Marianne might have thought Louisa fond of Miles. She wrinkled her brow. If Miles knew, he would stay away, that was sure – for a time at least. And she did not want her handsome son leaving her just yet to marry anyone. A love affair would be different. She had seen that Louisa Mortimer was not a girl to be dallied with. What had he finally told the girl? He had obviously been asked not to divulge his reasons for 'never seeing her again'. Poor boy, no wonder his letter had so many drafts. What did he feel for the child? He had never spoken of such intimate things to her and she would not have wanted him to.

She thought of her husband and a smile flitted over her fine mouth. Perhaps he had done little Marianne a service all those years ago. But what was to be done about Louisa? For once Kate looked irresolute. Miles must learn to aim for what he could have; he was better away. But Guy must not know about his daughter. That would be too tiresome. She hoped Miles would not take it upon himself, or perhaps one day be forced to blurt it out. One must keep one's dignity where one's children were concerned. She had been right, though, in her guesswork. Funny that it had never entered Guy's head. But of course men were rather stupid creatures. She was sorry for Miles. What could she do for him?

When René Boissier had returned to Paris he also found his thoughts turning now and again to Louisa Mortimer. He had not expected her to come back to his flat, and she had not. He knew she would breathe nothing to her mother of their meeting and his proposition. An ordinary girl might have collapsed into hysterics after a little reflection over what she had escaped. Not Louisa. Louisa would remain the girl he had not succeeded in winning. Still, he had had an entertainingly amorous two weeks with his old friend Rachel Quinn and things were better as they were. So long as no more Louisas refused him. He could not have that. To Sarah, who well knew of his little apartment, though he had never told her of it in so many words, he said nothing beyond, 'London was deadly dull – almost as bad as Paris in August. Let us go to stay with Maman in Cannes.'

Sarah knew this was the necessary prelude to being able to visit England at Christmas, so made her preparations for a visit south out of season. It was a nuisance, for Kate Demaine had invited her to a reception the following week, and she had been determined to go and observe Guy, in spite of her better self which told her to leave well alone. But she wanted to see him in his new guise now she knew that Louisa was his daughter. She sent a message to the Demaines explaining that she and René were off for a week or two and was surprised to receive a *pneu* back from Kate immediately.

> I shall be in the Hôtel Beau Rivage, Nice, when you are in Cannes. I would like you to call. Perhaps René would like to join a little yachting expedition with Guy that day? I shall fix the day later, so long as you can come. We have decided to forgo the reception and are off south tomorrow. Write there.

Sarah was mystified, but replied as requested. Kate did not want René to visit her. Now what could be behind that?

Louisa returned to London at the beginning of September to begin her 'helping out' in Tooley Street, travelling daily, partly by horse bus, partly by the new Underground, from Uncle Lechmere's tall, white house in its pleasant street in Hampstead. Great-Uncle Lechmere was as tall and pale as his house and not at all like her grandfather. He took a benevolent interest in her work, seemed to register no surprise that she should wish to do it, but talked to her instead after dinner each evening in a room whose walls were decorated with a William Morris paper. Really, if she had known how nice he was she would have come to stay with him earlier that summer!

'You are rather like your mother,' he had said, looking at her, she thought critically, on her first evening. Then he smiled and said, 'Off to work, then,' and disappeared into his study. Uncle Lechmere had no wife, but a housekeeper, a quiet, capable woman, who had a sitting-room and a bedroom on the ground floor of the house behind the dining-room. Louisa wondered why he needed such a large house for one person, but wondered no longer when Mrs Smith took her on a tour of the house. Every room, except his bedroom, into which she did not venture, and the dining-room, which she knew already, was piled floor to ceiling with books or papers or boxes of slides or cases of geological specimens. 'It's his work,' said Mrs Smith reverently. 'I dust when he's out at the Zoo or the College – he doesn't trust anyone else.' Louisa was intrigued, but got to know Lechmere no better after three weeks than after one. Yet he seemed to have a shrewd idea of her knowledge and opinions without taking any further interest in her, which she found a relief. Why did both René and Lechmere think she was like her mother?

The work at the Settlement was, however, not so pleasant or relaxing. When she entered the tall, drab building another world presented itself, a world in which every penny must be accounted for, a world where ragged women and their children came every night and where out of work youths congregated at midday to be dragooned into a football team or a table-tennis team by a posse of curates whose life's work was the eradication of the demon drink and the bringing to God of the repentant. She had at first been surprised that so much of the Settlement's work should be undertaken by men, but realised that her college was in fact only part of the charitable foundation and there were so many things women could not do. Ladies, usually well-educated ones, came in the afternoons to read to the children or teach the women plain sewing and the atmosphere was like a Sunday School. She read aloud herself as the women sewed, and was otherwise employed in the office to cast accounts and write letters. No one asked how she liked the work or why she was there and she felt unwanted and lonely, and berated herself for it. Certainly she did not want to stay bent over a little table in an uncomfortable chair in a room painted dark green for the rest of her life. She had not told her mother the work was unpaid and Marianne had been too sunk in gloom to ask. Louisa managed on the watery soup dispensed at midday and by the time work was over at six o'clock was ready to devour anything Mrs Smith felt like providing.

She knew she had made a mistake, and only her obstinacy, her feeling that she was offering something up to Miles, kept her going. She thought of Miles constantly, almost wrote to him once or twice, but was too frightened to receive his reply in case he cast her even further out of his life. One day though, after three weeks in which she had washed floors, wiped up vomit, accompanied a child to Guy's Hospital and barred the door to a drunken husband who came importuning his wife in the middle of the afternoon, she sat down and wrote a letter to her old Principal. She had done her best, but she was not cut out for this work. It horrified her. What must the world be like if the people who came to this were only a pool in a sea of misery?

She said to Lechmere that evening as they sat over their one cup of strong coffee, 'I don't think I am suited to good works, Uncle. But I don't know whether Mama really has enough money now that Papa is dead to go on paying my fees at Oxford.'

He looked at her mildly. 'Surely your Mama has explained her finances to you? What are they paying you at the Settlement?'

'Nothing.'

He raised his eyebrows.

'Well, I brought my savings with me and you have been kind enough to feed me – otherwise I couldn't have managed. You see they took it for granted at the interview that I would not be paid. Not even pin money. Ladies do it because they have social consciences and rich Papas. I didn't tell Mama.'

'Really? In that case *I* shall write to your mother. If there is any difficulty with the fees I shall lend you the money myself – no interest – capital returnable at the end of ten years. Will that do?'

'You are very kind,' she began. 'I didn't expect – '

'I know you did not. I expect your brother is being paid for, isn't he?'

'No, Hugh has left school and has gone to work with the firm in Manchester. Only Emma is at home.'

'You must go back to college, my dear,' he said. 'You could stay here, I suppose, and go to the Ladies' College in Regent's Park.' He looked enquiringly at her. 'Didn't you like Oxford?'

'Oh, yes, I was very happy.' She wished she could talk to this sympathetic scholar, but he would not understand. You could not imagine him young or in love. He seemed perfectly self-contained. She could not imagine that anyone so different from René Boissier could exist on the same planet.

'Well, decide what you want to do and I shall write to your Mama and to your Principal too if you wish. It's been a bad year for you, I expect?'

The Oxford term had not yet begun, so Louisa might return as if her employment had been only a little dabbling in good works at the end of the vacation. That made her feel ashamed.

'It's a drop in the ocean,' said Lechmere. 'You can't change social conditions unless you educate the people.'

'But it seems so unfair that people should have such dreadful lives in and out of the workhouse – women beaten, children hungry and men spending their time at the gin shop.'

'Yes, it is unfair. But the so-called "gentleman" spends a good deal of his time drinking, doesn't he?' he said drily. 'I think you should plan to teach in one of these girls' Grammar Schools – or would that not be "ladies work"?'

'I need to pass my examinations first – not that they will give us a degree, but I have two more years to do. I think I should go back to Somerville Hall, Uncle. Perhaps later I might do some further study here in London.'

'Right. Give 'em your notice at the Settlement. You could leave tomorrow if they're not paying you, but better give them the chance to suggest it.'

The Warden, however, was not surprised. 'We have another lady coming next week, a Miss Pennyfeather, a little older than you. Thank you for what you have done, Louisa. We'll always be glad to see you back.'

So that was another failure, thought Louisa. The only nice part of the Long Vac had been staying with Grace Wilkinson and her noisy family. At least she would see Grace and the others again if Mama agreed to allow her to borrow the money. She would be back in college no longer expecting to become a scholar, just a girl who needed her ignorance reduced. If she were never to see Miles again she might as well teach. Perhaps all men were fickle.

Louisa had, in fact, learnt something at the Settlement. She had learnt that she was not especially useful to the dockland poor who knew more about life than she did, and that she was no martyr. If she wanted to help the poor she might as well marry a curate. Armed with a letter from her uncle and permission from her mother to return to Somerville, she arrived in Oxford the next week on the first day of the new Michaelmas term, feeling older and sadder, but determined to make the most of her studies. She thought a good deal about her escape from both René Boissier and from the grind of uncongenial work and wrote that first evening to her mother to say that she intended to study hard. As for Uncle Lechmere, he was puzzled that his niece Marianne seemed to have little idea of what her daughter was doing. Perhaps she *was* short of cash and had therefore not pressed Louisa to continue her studies.

Marianne seemed to slip further and further into a combination of lethargy and heightened apprehension of a danger that existed all around her. She had had only one letter from Sarah to which she had replied at first with only one word, 'Thank you,' and she was beginning to wonder whether Louisa might one day guess the truth. Might that not relieve her own dread and sorrow? But she could not make up her mind to tell her herself and was grateful only that she need not see her daughter until the end of term at Christmas. She had been neglectful. It was kind of Lechmere to offer to lend the money to make Louisa more independent. The probate was still not settled on Ned's estate.

While Marianne was spending her days dragging herself from her couch to the garden, restlessly devoured by the spectre of her own sins and guilts, she did find just enough energy to write again to Sarah, but could not seem to express herself clearly.

Sarah was also suffering a good deal from her recent knowledge of René's latest *affaire*. He had taken care she did not know about his meeting Louisa – but he had not been too careful about letting drop that in London he had seen an old 'friend'. So she looked forward to Kate's invitation to spend a day in Nice with her. It would be a relief if René went off for the day with his old friend Guy. She often longed for solitude now her children were back at school in Paris.

It was with a certain curiosity, mixed with apprehension, that Sarah arrived at the Hôtel Beau Rivage, a small but highly exclusive, establishment frequented by rich English and Americans, with the best chef on the Côte and the cleanest and most tastefully furnished rooms. Good taste was expensive, thought Sarah, as she was ushered into the private sitting-room that looked out over the bay.

Kate was wearing more rouge than Sarah had ever seen before on her alabaster cheeks and looked a little strained when she entered the sitting-room from her boudoir. There was no evidence of the fact that a husband shared the apartment. Sarah wondered if he had his own suite of rooms. Mrs Rhodes, Kate's new American maid, had withdrawn after showing Sarah in and Sarah took Kate's outstretched hand, which seemed to waver a little in her own palm.

'I sent for coffee,' Kate said and pointed to a little lace-covered table in an alcove where a coffee-pot was steaming away on a silver tray. 'You will have some?'

Sarah took the porcelain cup on its plate and the snowy napkin that was proffered and did not seat herself at first, but took the cup to the window and looked out over the silver sea. It was a beautiful room with a balcony curving out from both sitting-room and boudoir, the curtain tassels holding back a double drape of lace and velvet. René would appreciate such a suite. He ought to have married Kate, she thought, with a slight stab of envy. The coffee was also delicious and she sipped it and then turned to Kate, who said, 'You are looking well, Sarah.'

Sarah coughed a little nervously and then came to sit beside her friend who was attired in an ivory skirt and pale coffee-coloured blouse and a long string of pearls. She gathered her wits as Kate poured a stream of fragrant coffee for herself, seemingly reluctant to say whatever it was she had summoned Sarah to hear. She put the cup down and waited, hands folded.

Kate stirred her coffee and then said, 'I want to talk about Louisa Mortimer and my son.'

Sarah's heart gave a jump, but she said nothing. Perhaps Miles had told his mother of her own talk with him but she did not really believe he would have done.

Kate went on. 'You are a good woman, Sarah,' and then stopped. Sarah felt embarrased. Kate seemed to have relapsed into thought for a moment, so Sarah said rather nervously, '*I* am not "good", Kate. Why should you say that? No one is really good, except saints and horses.' She smiled. She had made a joke and the tension was alleviated. Sarah had the feeling that somehow she had the moral authority but could not imagine why, nor that Kate would continue to approve of her if she knew that she had been the go-between for Marianne with her message to Miles.

Kate took a pensive sip of her coffee and then said, 'The men will be away for some time. We shall not be disturbed. Did you know then that my husband seduced Marianne before her marriage?' she said quietly, without looking at her friend.

'Yes.' Sarah blushed. But how did Kate know?

'Never mind how I know,' said Kate. 'Did Marianne tell you?'

'Only after Ned's death last February – in great distress. I'd never guessed such a thing. Poor Marianne! She was going out of her mind, I think. And poor little Louisa.' Sarah looked to see what effect the last three words had had and Kate blinked slightly when she saw her gaze.

'Yes – well – I gather Louisa and my son were becoming fond of each other last year. Was that why Marianne told you?'

'They might have begun to love each other,' Sarah said.

' "Love" in the way Marianne used to speak of it? I have never experienced it myself,' said Kate coolly. 'A little tenderness, a little excitement?'

'Have you asked your son about his feelings?'

'Oh, no, I could not do that – ' Kate seemed embarrassed. 'I was going to ask you what *you* thought. After all he stayed with you when Louisa was at Lake Hall Park – I had no idea that he was writing to her even.'

'How do you know, then, if he has said nothing?' asked Sarah directly.

'Never mind, I always suspected though. But you must believe me. When I married Guy, I had no idea – when he asked me to marry him – no idea that . . . I would not have believed it then anyway! No idea in the *slightest*, not even a suspicion, that he had been Marianne's lover. I would have let him marry her if I had known.' Kate, the invincible Kate, was agitated.

'Oh, he wouldn't have married Marianne,' cried Sarah, before she could stop herself. 'She had no fortune and no power over him. She has paid for her mistake, and so will Louisa.'

'Then it *is* true that Louisa loves my son?'

'Perhaps,' said Sarah. 'But he has gone away.'

'She will get over it!' said Kate.

'But will *he?* Miles is a serious young man, you know – ' Sarah stopped. Kate seemed to be lost in thought for another long moment. Sarah waited, unsure what Kate wanted, or wanted to hear.

'Ned never knew about Guy and Marianne, then?' was Kate's

next question. She stood up and looked out of the window herself before turning again to Sarah.

'No, she never told him. She could not bear to – was perhaps afraid to – after all that deception. She thought she was doing the best thing in accepting Ned when he asked her to marry him. She knew already that she was expecting Guy's child. And Louisa was always Ned's favourite child. Marianne says he could not have borne it. It would have destroyed their marriage – they *were* happy, you know.'

'Now he is dead she need not worry about him,' said Kate rather unkindly.

'It is Louisa she is frantic about – that Louisa and Miles might want to marry – she never wanted them even to meet! It had to be nipped in the bud.'

'So she asked *you* to tell him, did she? Poor Sarah. What a mission.'

'Yes. I told your Miles that he had a half sister and that he must not see her again. I advised him to go away, in case he might entertain feelings for her that were not sisterly ones. I'm sure he was fond of her. Did he say nothing to you?'

'No, he avoided me and his father before he went to the States. I expect he is rather unhappy if what you say is true.'

'He is very shocked,' said Sarah. 'Not the sort of thing to hear about your father at his age.'

'I don't think that would be the worst of it. Miles has little in common with Guy, but he is used to Continental *moeurs*.'

'I'm sure he does love Lulu,' said Sarah sentimentally.

'I think it could blow over,' said Kate. 'He will forget her in the end.'

'I wish that were true. He had just begun to think seriously about her, I think. But what could I do? I might not have been too late, but it is Louisa I am sorry for as she has no idea of the truth.'

'Louisa is paying for her mother's past falsehoods,' said Kate.

'If it had been you, you would have told your daughter who her real father was?'

Kate was silent. 'I need a drink,' she said and rang the bell. When Mrs Rhodes came in from a room down the corridor, Kate ordered two glasses of wine and when they were brought in, handed one to Sarah with a slight moue. Sarah noticed the bracelets glittering on Kate's neat wrists and the rings on her hands and waited. Kate did not seem sorry for Louisa in the way she was. They sipped their drinks. Sarah felt the colour rising in her cheeks. Kate remained pale, except for two high spots of rouge and lay back in her chair, her foot tapping on the carpet. 'I must confess that I don't want my son to marry young. Too many marriages are made in the rashness of youth. He must be free to be himself and to carve out a career. I always wanted to be a doctor myself, you know. If I had been poorer and a man, that is what I should have done. I wanted to be a surgeon. Miles will do all I could not.'

Sarah wondered if Kate regarded her own marriage as a mistake, but knew her friend would never confess directly to failure. What kind of marriage did Guy and Kate have? She was puzzled, had always been puzzled over Kate's real feelings.

Kate went on. 'He is a very clever boy, you know – and very steady.'

'But if they had begun to love each other – to envisage a future together – it would be just as much a blow to him as to Louisa. Even when he knew the truth . . .'

Kate swallowed her drink and put the glass down on the table. 'Can you think, Sarah, of anything to be done to improve matters? It won't undo what Guy – what my husband did, or relieve Marianne of the guilt that is obviously devouring her. The question is, who is the most important in all this? If Marianne had kept her peccadilloes to herself my son would have probably in the course of nature stopped feeling whatever it was he thought he might feel for Louisa Mortimer.'

'It is Louisa whose heart will be broken. I wanted Marianne to tell her, but she could not. She seemed sure of Louisa's feelings. She knows her own daughter.'

'Isn't Louisa bound to find out one day? I don't want Guy to know. The idea that he might be Louisa's father has never occurred to him. It's far too late for that.'

'If Miles did see Louisa again – and told her – would that not be better for her?' said Sarah. 'As it is she will put his cooling off to a natural male reaction. Of course, on the other hand, if she knew the truth it might frighten her and blight her future with others. It would reduce her mother in her eyes – and put a barrier between herself and her natural feelings.' After this rather long speech Sarah looked away.

'How well you express it,' said Kate. 'She might also decide to leave the paths of virtue.' Sarah said nothing. Kate went on, as if to herself, 'If only I knew what Miles really feels. We may be assuming that they would have eventually decided to cast in their lot with each other when it was only a youthful passion – gone as quickly as it started. Opposition might lead them to think they are in love, when a long period away from each other with nothing more said would effectively stop its progress.'

'Your son is an honourable young man,' said Sarah finally. 'He will not tell her the truth, but neither will he put himself in any situation where he might be tempted to allow his feelings full rein. I am sorry for him. The things we feel when we are young mark us for ever.' Kate looked at Sarah and wondered whether she had regrets at the way her own marriage had turned out. There were all sorts of stories about René Boissier and his *amours*. Poor Sarah. Poor Marianne. She herself was the only one of the three not to be pitied!

'So you think Louisa would have been willing to become his lover if she thought he loved her? Like her mother? Throw herself at a man?' asked Kate cruelly.

'That was lack of experience. Marianne thought Guy was serious about her, I am sure.' Sarah sprang to her friend's defence.

'She had no *savoir vivre*. It was unusual though, was it not?'

'Marianne was besotted with Guy. She was shattered when he left Lake Hall Park and married you not long after.'

'Till Ned came riding up on his charger to save her. I know Guy. It meant nothing to him. He thought all English girls were fair game. Louisa, the result of an accident! – that isn't very pretty either, is it? Marianne should have had more common sense. It was not as if *he* could help himself. She tempted him and did not even realise that she was doing so. Thought it was a great love. Perhaps Louisa is the same. How stupid young women are.'

'Oh, no, she is *not*,' cried Sarah, distressed. 'She and Miles do like each other, you know, apart from anything else. They have corresponded with each other for some time – after they met in Paris at our house, and he was most desirous of seeing her again. It was at Lake Hall Park they began to feel something for each other, I believe. So it is partly my fault for inviting them together.'

'They know each other quite well. That's a pity.' Kate's cynicism was lost on Sarah.

'Yes, I think so. I can't imagine what Louisa must think has happened. It is too cruel.'

'I think I shall have to speak to Miles,' murmured Kate. 'It depends on their true feelings.'

Sarah looked puzzled. 'To comfort him for her loss?'

'If it is not too late,' replied Kate enigmatically. 'If only Marianne had come to me, not to you. But she never trusted me, did she? Perhaps she was right not to. Still – ' she paused,recollected herself, then in her best *grande dame* manner, said, hand outstretched, 'It was good of you to come, Sarah. I may go to see Marianne one day.' Sarah looked aghast. 'Don't worry, I shan't say we had this conversation – you have been very loyal to her.'

As Sarah put on her gloves, Kate said, 'I think I've had enough of Europe. I expect Guy and I may settle in Boston or New York when I have Kitty safely married. But I'll be over in London soon. Would Marianne ever come to see me in London, do you think?'

'She never moves from the country now. Since Ned's death she says she has felt strange. She is ill, I think and lies on her sofa, I expect, as she was lying when I last saw her. Oh she was dreadfully tormented. Almost as though *her* life as well as Louisa's was ruined. I don't see what you could do for her.'

Kate distractedly bit some loose skin from her little finger, then put a hand up to tidy her hair. 'I *shall* go and see her – when I have seen Miles. But she should have told Louisa the truth as soon as Ned died. It's always better to tell the truth. Louisa might have borne it better than my son.'

Sarah looked into the cool grey eyes under the fashionably arched brows.

'You can pray for the two of them,' added Kate as Sarah went out.

The 'interview' was over and Sarah had not had the courage to ask Kate's advice over her own husband. Only as she walked out of the hotel and down the palm-fringed avenue du Casino where the carriages were lined up did it occur to Sarah to wonder why exactly Kate had wished to see her. She had seemed concerned, of course, that her son might be unhappy; less concerned for Louisa and not at all surprised at Marianne and Guy's past history, but there was something else, something which Sarah was sure that Kate was not telling her. She wondered what Kate had ever seen in Guy Demaine. Perhaps it was better that she had said nothing about René. Kate was cynical and would have said that Sarah should have known what she was letting herself in for years ago.

21

Kate Demaine, *neé* Mesure, had been neglected as a child by a rich mother who was too busy enjoying herself to spend much time with her little daughter. Left early in her marriage by a husband who had later been killed while living a profligate life out west, Caroline Mesure herself had succumbed to typhoid fever, leaving her brother to bring up the little Kate. Kate had never known stability except with Uncle Adolphe and his wife, but had learnt early to disguise her real feelings and to erect strong defences around her private space. Yet she did not lack courage or sensitivity. She often wished she could be more open in her dealings with others, but suffered from a dreadful paralysing inability to appear vulnerable. It was so much easier and pleasanter for others if you expected nothing from them and gave the impression of not needing their sympathy. Guy's insouciance had, in fact, begun to worry her. It was all very well having a complaisant husband, but Guy could not be relied upon to take over responsibilities except in the matter of practicalities. His parties had made her famous and she commanded an expensive circle of acquaintances who saw Guy and herself as an eminently suited couple. She had even breached the last bastion of the Boulevard St Germain. She had enjoyed herself throughout the last twenty-five years in ways which suited her nature. She liked conversation, good wines, amusing company, the purchase of beautiful clothes, the attention of handsome men – but there was an emptiness at the heart of her life which no one except perhaps her son Miles guessed existed.

Miles, what was she to do about him? She loved her son, though remaining amazed that such a stable, gifted boy should have been born to her. She could not bear that he might be really unhappy or that she might not in some way remove any obstacle to his peace of mind if it were at no cost to herself. But her early childhood made it difficult for her to contemplate giving up anything so painfully acquired, and whispered to her that if she had managed to make a life despite her unsatisfactory parents, aided of course by money, then Miles too must learn to put up with what he could not have. She resolved to find out from Miles himself the extent of his feelings for Louisa, but she would have to be tactful; she hoped that Sarah had exaggerated. If not, she would go to see Marianne.

Marianne, for her part, was still in a strange depression which made her feet like lead and her mind an alternation between a similar thick heaviness and an equally unpleasant wild overstimulation of the nervous system which also precluded rational thought. Yet through the gloom and the jumpy unease, the feeling was growing that she must in the end confront her daughter. She did not fear that Louisa would judge her too harshly, but she shuddered to present an example to her of a mother who had disgraced herself and also to change her daughter's happy memories of her father. Ned had loved her and she had loved him. What right had she to take away the child's past? Louisa might wish to meet her real father and that simply could not be allowed to happen. Apart from anything else she knew that Guy had cared not one fig for herself and certainly did not have the right to lay claim to a beautiful daughter whom he had never known and probably would not have worried about if he had. In the midst of her nervous prostration and the fears that stalked through her dreams and her waking life and made a mockery of everything in her present, Marianne was also feeling lonely. She had allowed Louisa to leave Oxford, as she had thought, only to have the child change her mind. What was to become of her? The Manchester business was not doing too well and Louisa had no prospect of marriage on the horizon. She would like too to be able to talk to Emma or Hugh about their finances and futures, but Hugh, who was learning the business, brushed aside any overtures from her and said everything would be fine in the end.

Emma had her own private domestic life, could be relied upon in practical ways, but was too young and shy and unworldly to understand that life was a perilous business. Marianne knew that she had had the opportunity twice now in her life to show moral courage and each time she had failed herself. Her existence now seemed pointless, except to continue as a mother, as a widow, until such time as her children should take the reins of domestic government from her. The only things which could bring a moment's cheer or interest to her wretched life were the writings of the poets and even they were not much help. She felt so lethargic. Yet she still tried to read, hoping that some writer sage would one day come to her aid and tell her what kind of person she ought to try to be. The death of Matthew Arnold, whose writings she venerated, had even added to her dreadful grief for Ned in a strange way. That grief was a relief from her overpowering guilt. At other times she told herself she was not fit to mourn Ned, that her life had been built on a lie. She tried to read French to take her mind off her troubles and had on her bedside table the *Vie de Jésus* of Renan, a book she knew was shocking to the right-minded. But she could not make the effort. Again and again she told herself she should have had the courage to say what she really felt years ago, to have run away abroad with her baby and to have refused Ned's offer of marriage. She saw it clearly now. But even at the worst moments of her depression, she could not truly say that she had not been a good wife to Ned. Poor, dear, Ned, who lay alone now in the churchyard never to know the burden of her guilt and, therefore, making her feel doubly guilty. She even thought of suicide. But that would leave her children even more bereft. Her duty was to them. She must carry on alone in her awful pilgrimage and not look forward to anything better than eventual extinction with no forgiveness from the god in whom she did not believe.

Her children, of course, had no idea of the torments of their poor mother and thought only that she was grieving for their father. They missed him too. Hugh tried to follow him as a capable businessman, but was none too sure of his own abilities, even if he gave the impression of ambition and premature yielding to convention. Emma knew only that her mother was unhappy and that somehow it was connected with Louisa who was always away now. She had never been close to her sister and felt even more remote from her as she lived her quiet Langthwaite life, attending church regularly, gardening, and looking after her little flock of fowl. 'Miss Emma's a throw back to her great-granny,' the villagers said, and it pleased Marianne to have provided at least one satisfied ordinary practical child to carry on Ned's family characteristics. But she did not understand this daughter who was closer to her son Hugh than she was to her mother. It was Louisa whom Marianne loved passionately and mourned as though she too were dead. Not another word had passed between them on the subject of Miles, but Marianne had noted, with sinking heart on Louisa's visits home, that the girl's vitality seemed to have been quenched.

As for Sarah in these sad days, she suffered once more in silence when she returned to a Paris of dazzle and opulence and saw that her husband would never change his habits. She devoted herself to charity work, was tolerated rather than accepted by the Catholic Sisters of the *quartier* and continued to dream one day of returning home when Valéry and Olivia were settled in either marriage or a profession. With René she was almost shy now. They ran the household in their separate ways, neither quarrelling nor entering upon any conversation more intimate than a discussion of household accounts.

But events were to disturb the illusion of an even tenor in the lives of Sarah, of Kate and of Marianne, events which bypassed Guy and René, who were the causes of unhappiness to all three of them in their different ways.

Kate Demaine began the sequence of events by writing to her son Miles and asking him to return to Europe to see her. It was now the summer of 1889. Guy was gambling pleasantly in Monte Carlo which suited Kate, who wanted him well out of her way. Also the Prince, of whose son she had hopes for Kitty, happened to be on the Côte, so Guy had taken his daughter with him to impress Vrigne. Kate was not sure whether Guy was quite up to carrying out her plans for Kitty, but at present she had a much greater problem to solve. Kitty, not averse to enjoying herself and equally aware that her Papa was too lazy to keep an eye on her, had other plans.

Now nearly seventeen, and a beauty, Kitty would amuse herself flirting with various young men who thought she was much older than she was, but she had not yet met anyone to whom she might give her heart for safekeeping.

Sarah went to Lake Hall Park in the early summer. Lady Penelope had had another stroke and was not expected to survive it. René was pleased to stay in Paris and enjoy to the full his city and his women. Paris had reached the apogee of its splendour and the

English Prince of Wales carried on there in much the same way as René.

Louisa was about to finish her second year in Oxford. She had become thinner and quieter and was unsure even now whether to continue her studies. She read much in Swinburne and wept over him while at the same time realising that Miles was worth more than a few sentimental tears. But Swinburne's 'Time with a gift of tears' went to her heart and gave her the illusion of having at least lived and loved. There were many young men in the town who noticed her as she went about with her friends to lectures and libraries, but she had no eyes for them. The sight of any man who was not Miles Demaine now made her feel physically ill. During the year she had tried religion, frequenting an Anglo-Catholic church which was lavish both with bells and smells, but the church did not seem to have the answer for her broken heart. She was not the stuff out of which nuns were made, though for a few days she was possessed with the idea of committing her soul to God and renouncing the love of men. But the thought of Miles returned and became the constant accompaniment to her inner life and her studies. For relief, like her mother, she turned again to books, tried to read William Morris and wished she had a mind that would enable her to see beyond the merely personal to a bright vision of society. His 'class against mass' appealed to her and was even the subject of a letter to her Great-Uncle Lechmere. That kind gentleman invited her to visit him again in Hampstead, and took her to see the first performance of Ibsen's *The Doll's House*, which alternately exhilarated and depressed her when she thought it over. If Ibsen were right, she had been saved from a life of feminine imprisonment, but she refused to believe that all married life was like that. Her own mother and father had been happy, she thought.

She returned to Westmorland in June to find her mother just as sunk in misery as ever. The first conversation she had with her was a stilted one in which she tried to interest Marianne in her reading, and Marianne, who usually could be relied upon to discuss books rationally, seemed far away. Louisa still did not mention Miles, and Marianne was relieved. Louisa for her part tried hard to understand her mother, even though she was sure now that she was behind Miles's flight to America. Just for one moment she thought that she had got through to her when, reading Morris on work and rest, she quoted his words about rest being useless if it were disturbed by anxiety. Looking up at Marianne she said, 'That is true, Mama, I feel you do not sleep well and are troubled – I wish I could help you.' But Marianne went pale and muttered something about the young not understanding how dreary the world was and how dangerous, and Louisa held her tongue.

Just at the time that Miles Demaine was returning home to his mother, Louisa was trying to cheer up her own female parent. She would hear Marianne tossing and turning in her bed when she passed the small room where her mother now slept. She had refused to sleep in the big bed where for so long she had lain with Ned and sometimes when Louisa lay awake herself she would hear or imagine that she heard her mother crying out or groaning, though whether she was asleep Louisa was not sure.

'Does she often cry out at night?' she asked her sister.

But Emma did not know, slept so well herself that she had never heard anything. Hugh was kept busy most of the time in Manchester and was in any case embarrassed at his mother's protracted mourning, so could not be approached.

One night Louisa got up at about three, having woken out of a dreamless sleep as though someone had called her. She was instantly awake and decided to go down to look at the lake which was lying still in the moonlight. The servants were all asleep on the top floor of the old house and Louisa crept downstairs, intending first to drink a cup of water. Passing her mother's room she again heard the voice and stopped for a while at the closed door wondering whether she should go in and comfort her. But she was shy of her these days and still angry with her. She had done her best. Mama did not seem to want to feel better. Even Mr Carmichael said that he was worried about her.

Just then Louisa heard her father's name cried out from behind the bedroom door. Was her mother awake or asleep, or perhaps in some twilight world? She put her candle down on the floor and leaned against the door embrasure to listen, feeling it was almost a duty to eavesdrop. She heard 'Oh, Ned! Ned. Forgive me, Ned' repeated quite clearly, and then a silence. Then there was weeping. Louisa moved away but was held then by the sound of groaning that followed. She stood fascinated, yet irresolute, wanting to go in and comfort her mother, feeling an almost physical need to put her arms round her. When there had been silence for five minutes she took up the candle to continue downstairs. Then she heard again a low moaning and a 'Forgive me, forgive me' that seemed to be tugged out of her mother's body in agony. What was the matter with her? People did not always ask for forgiveness, did they, when someone they loved died? It was true that at first she herself had felt that she had not been nice enough to her father, but that feeling had gone. It had been in essence a selfish feeling, a luxuriating in agony. What her mother was saying was somehow different. He mother and father had hardly ever quarrelled. What did he have to forgive her for – ? What did she *think* she had done?

Disturbed and uneasy, Louisa waited a little longer, but Marianne must have fallen asleep. Louisa went back to bed without drinking her cup of water or going to look at the lake. Instead she lay awake trying to understand her mother and wondering what it was all about. She had seemed to be pleading with her dead husband for forgiveness, had sounded rational, not delirious, in spite of the crying – rational and despairing. All she could think of was that her Mama had been unfaithful to her Papa – but the very thought was ridiculous! Louisa felt afraid. The next afternoon she plucked up her courage and said, 'Mama, Papa would not have wanted you to grieve so.' But Marianne looked at her with a face so full of what Louisa could only describe to herself as terror that she said no more. Why did Mama look at her nowadays as though she were afraid of her own daughter? She had never seen her look like that before.

Miles Demaine returned to Paris to find his mother alone in the large apartment for Guy was still on the Côte with Kitty. The Prince had left Cannes once the yachting was over but Guy had announced that Kitty preferred to stay with him for a time. Kate had been a little puzzled and thought that the casinos must be the reason. Perhaps he had really gone too far this time and wanted to stay to recoup his losses. It was a pity that the German casinos had been closed these last fifteen years. She had always preferred Guy to gamble there. He mentioned that he had seen the grass-widower René Boissier who was there paying a visit to his mother. So Sarah would be in England, thought Kate, who had no particular desire to see her friend until after she had decided what to do, and done it, and perhaps not even then. She wished she could go and rescue Kitty, for she did not trust Guy to be strict enough in his household down there, but there were other more important things on her mind.

Her interview with her son was short and to the point, but afterwards she felt exhausted and a little lightheaded. She had done the right thing.

'Yes, I do love Louisa Mortimer,' he had replied to her first question. To his: 'Was what Aunt Sarah said to me true?' she replied with a curt 'Yes,' and a 'Do not on any account tell your father – he does not know and shall not.'

When Miles murmured something like 'Louisa should know, Mama,' Kate waited a moment before replying. Then she replied, 'Leave me to worry about that. So long as your father never knows. I am going to see Louisa's mother in England.'

'So at least Louisa and I can be friends?' he cried, grasping at a straw.

She ignored him and went on. 'Will you go to Lake Hall Park when I telegraph you? I can't explain now. Will you?'

He shook his head in puzzlement. 'Will you see Louisa, then?' he asked.

'Leave that to me,' she said again. She decided though to enlighten him further and after a pause went on. 'I have written to Sarah. She will receive you there. Marianne will not be told that you have been invited and should have no objection therefore to her daughter paying Sarah a visit. I've asked Sarah to invite her there now.'

'Is that quite fair?' he objected. 'How can I act naturally towards her when I know what she does not?'

'Do as I ask,' was all she said and she turned the subject once more to her husband. 'He'll stay on the Côte with Kitty for the time being,' she said. 'You won't see him till you return from England.'

Without saying it in so many words, Miles had known that his mother found Guy an increasing problem. Her own investments were secure, but business was not doing well anywhere in Europe. For the first time he heard his mother state that she wished to settle back home in the States where the gambling was not so much of a problem, except out West. 'What if Papa does not wish to go there?' he asked her. He saw that his mother was more vulnerable than he had ever imagined. Up to now she had always managed Guy.

'He always goes where I want. He has no money of his own,' she said finally.

'But that is not entirely his fault,' said Miles. Kate accepted the implied rebuke. Her son looked sometimes almost a stranger to her, yet she knew she could rely on him. He had told her of his feelings for Louisa and she had had to conquer jealousy – jealousy over her own son!

Miles could scarcely believe that he was going to see Louisa again. The year that had passed had been the most painful and dreary of his life. If it had not been for the fact that, in spite of his misery, he had worked hard at his studies, he did not know what he would have done. Kate was aware of all this. Meanwhile she was making arrangements to travel with her maid to England. He thought, Mother is playing some sort of game.

Sarah was disturbed to receive Kate's letter and had no one in whom she might confide. She was to invite Louisa, and keep from Marianne that Miles was later to be her guest. That was against her nature and she could not imagine what Kate thought she was doing. 'Believe me, this is the best way. Trust me,' Kate had written.

Sarah then was to be chaperone – but for how long? And when would she have an explanation? But Sarah did what she had been asked and wrote to Louisa on Kate's instructions and suggested that she might vary her summer vacation by coaching Olivia in English written work. Olivia's spoken English was adequate – after all, she had an English mother – but her written English was appalling, and if she was to fulfil her ambition of becoming a businesswoman and founding her own fashion house one day, she must improve it. The richest clients were always Americans, along with English Royals. Sarah knew Olivia was now determined to be a designer.

Louisa therefore journeyed to Lake Hall Park with no other thought than how kind it was of Aunt Sarah to find her an occupation and release her from attendance upon her Mama, who was growing stranger day by day. Marianne had made no objection.

Sarah and Olivia were in the morning room where Louisa had been ushered by a parlour-maid. All Louisa could think of at that precise moment was how Aunt Sarah's husband had invited her to become his mistress. As she greeted Sarah she thought, It is absurd. I don't want to have secrets from Sarah, but this is something I can never tell her. Yet Sarah was still the same person she had always been. She could not act differently towards her. Sarah did not seem to notice anything and some of the constraint on Louisa's part was soon loosened by Sarah's dog bounding up and by Olivia's taking her arm in a very friendly way. More friendly than she had ever been in fact.

There was a tray with glasses and a decanter of Madeira on a little table. Sarah obviously felt that Louisa was now grown up and she was poured a drink by the maid which took away the rest of any awkwardness she was feeling. She made a joke about being the 'governess' now and they all laughed. Provided René was not staying at the Hall. Louisa began to worry again about this till Sarah said, 'My husband is with his mother in France. She is worse, I fear, but she has a strong constitution. Unlike my own poor mother.' Lady Gibbs was now bedridden.

Afterwards, Louisa and Olivia went for a walk in the rose garden and again she had felt strange when she thought that this girl's father had made her a proposition. She wondered whether Olivia knew much about her father's nature.

'I fear the roses are much reduced and need attention,' said Olivia with a sigh.

'I shall prune them,' said Louisa. 'My mother has wonderful roses too.' She thought of Marianne again, but turned her thoughts away from her and began to sketch her idea for improving Olivia's English. If she were the 'governess' she had better be an efficient one. 'I thought you might write about your interest in fashion and what you want to do with your life,' she said rather shyly as they sat on a rustic seat.

'It's kind of you to take such interest in me,' said Olivia. 'We had such boring English lessons at the convent – you know, bits of Shakespeare to translate, and Sheridan. I understood them but I couldn't put them into French. They weren't *meant* for French. I want to know the English words for business and fashions and even stitching. Mama doesn't know them all – *you* will be able to teach me!'

'I think it is you who will teach me,' replied Louisa, trying not to envisage vistas of boredom. 'I have brought my big dictionary and if that won't do you will have to draw things for me and I'll do my best. I could perhaps improve your spelling and written English – ' She began to warm to her subject and the unemotional Olivia even squeezed her hand.

'I'm so glad you came,' she said. 'I get so bored here.'

'Lessons every morning and conversation or reading aloud in the afternoon,' said Louisa smiling, and they went into luncheon better friends than they had ever managed to be previously.

The days were spent more or less as Louisa planned, though Olivia's span of attention was not long. Louisa found herself wanting to educate her in matters other than sewing and fashion, but tried not to appear to missionary-like in her efforts. In the evening they both played the piano and Sarah joined them for a game of cards. It was a pleasant, if undemanding, existence – less fraught than her days at Langthwaite now were. But she could not help feeling that her life had come to a full stop.

The Prince had returned to Paris and paid a private call on Kate. It was the day before she was to leave for England and most inconvenient. Miles was fortunately out with a male friend. Still, with an eye on Kitty's future she was all amiability, though she knew he was more interested in her than marrying his son off to her daughter.

Kate had never allowed any physical intimacy with her 'lovers'. Recently she had even tired of male admirers. What she had enjoyed was the sense of power; it seemed somehow to compensate for the miseries inflicted upon women in so many ways by men. But now this had begun to pall. She put it down to her age, but was also aware that she had changed. It depressed her rather that there was nothing to look forward to. When she had told Sarah that she would like to have been a doctor this was partly true. A surgeon also had power, as did even a humble physician. But enjoying herself, which she had never before found difficult, had been more important than work, not that anyone in her position – even a man – would have needed to do that. The only thing left to her, she sometimes thought, was to spend money. And spending money meant shopping. Shopping for pleasure could still raise her spirits. Accordingly, when the Prince had gone she took the carriage and went to her favourite spot on the rue Royale where she had ordered what she carefully calculated would not be too obviously French for England – a dove grey costume with pearl buttons which looked deceptively simple but whose tailored perfection set off to the utmost her skin and eyes and hair. A grey hat with a fine veil and a pair of grey gloves and suede boots and parasol completed the outfit. Then she returned home and sat for a long time before her desk.

She was not a great reader, except in the columns of newspapers and financial journals, and neither did she particularly enjoy music. She did however indulge in some rather fine crewelwork and her tapestries were quite exquisite. This evening though, when she had balanced her accounts, she sat doing nothing, thinking of the task ahead. She shrank from it inwardly. She thought she cared little for the opinions of others, but did care for her son's good opinion. It was a miracle that he had turned out as he had. Still, one must be cautious. Heaven knew what would happen after her mission to Marianne. Thank God Guy was still away. She had enough thinking to do at present and his incessant search for worldly pleasure would have got in the way. He would want her to resume her parties and receptions. But once Kitty was settled she could stop trailing about and do something quite different.

On her return from England she must write to Uncle Adolphe, still alive – and always alive to her best interests. She would set herself up in Boston and become a legend for elegance and discriminating charity, she decided. She would also buy a farm-house in New England where she might retire whenever she was tired of town. There might even be compensations in new acquaint-ances. Frenchmen were all very well, but somehow one could not take them too seriously. What she needed was an *amitié amoureuse* with some kind American gentleman who would adore her and admire her in return for her interest in his work. Provided Guy had enough cash he would fend for himself, might even enjoy flirting with American matrons who would appreciate his Continen-tal good looks and charm. Kate was aware that she used people. Why not? Did they not use her? She was cynical about herself and humankind in general, but also proud. What she had to do pre-sently was to humble herself. She gave a grimace and checked the *papier poudre* on her nose and chin before ringing for a light meal. She would have to steel herself and be brave. That was one quality she had not usually lacked.

Kate knew she had taken a gamble with her son in telling him what she had. Miles however would be discreet. He had inherited enough of her own brains to know that Guy must never know about Louisa. Not only that Guy might be insulted or even amused, but that he might embarrass poor Marianne in her lakeland hide-out. She was sure Marianne had no wish to renew her acquaintance with him. Yet she had seen some of her husband in Louisa, a quality of quick sensuality, suitably attenuated of course in a well brought up English Miss. Provided the child herself did not feel she must meet her father. But there were so many 'provideds'. Kate put it all out of her mind for the night and settled to sleep. She would be up early, take her maid Marie with her on the channel crossing and then leave her in London at a suitable hotel for a night or two. By the time they were back in Paris matters would be out of her hands.

Marianne held Kate's telegram from London in a trembling hand. Everything made her tremble nowadays. What could Kate possibly want with her? Did she know about Guy and her? Was she coming to wreak vengeance? Kate had never seemed especially interested in her. Perhaps it was about Miles. Perhaps Kate had decided Miles and Louisa were suited and wanted to discuss a possible future match. She knew nothing whatever about Guy's own past and wanted Louisa for a daughter-in-law? But that was ridiculous. Louisa had not mentioned Miles for a year and Miles was, she thought, in America. What could it be? All the telegram said was 'Arriving Royal Hotel Ambleside and will be with you for a talk on Thursday 4 p.m. – Kate Mesure-Demaine.'

'For a talk.' It sounded sinister. Marianne realised that she was rather frightened of her old friend. If only Ned were there, she found herself thinking. But even Ned, who had usually been tolerant, had clearly never liked Kate and had pointedly avoided her, she remembered, even when she and the children had gone for that Christmas – when she herself had also been frightened – at Lake Hall Park. Yet it had not turned him away from Miles when she had thought it would.

It was a beautiful day. Marianne went into her garden and looked up at the line of hills in the north whose distant peaks always made her feel that over there, somewhere, was a magic land. For one second she had forgotten her misery, but it returned, as it always did. If that half-hour of her life with Guy Demaine had never happened . . . if Ned had not died . . . if Louisa had never met

Miles, or if, having met him, he had taken no interest in her or she in him . . . Marianne felt, as she often did, that nothing was fair. She had once, in the distant past, wanted Guy Demaine. Even now she was not exactly ashamed of those old feelings. But why should half an hour be paid for in years of never-ending misery? Would it not be better if she just slipped out of life? Yet Louisa, her darling Louisa, must be happy, must not pay for her own sins. She had, in spite of her anxiety about her, neglected her recently, not taken enough interest in her studies and her feelings, so consumed with terror she had become. And now even Louisa had deserted her.

It was half-past three. Soon she would have to confront what she felt would be another messenger involved in her destruction. She felt tired. The attempt to walk for a few minutes in the garden had sapped her of her poor energy. Perhaps she would die; perhaps some dreadful disease had even now reached into her. She was helpless; there was nothing she could do but wait.

22

Kate Demaine was getting ready to walk over to Langthwaite from her hotel. She did not usually walk but had felt a sudden surge of animal spirits. She would think over what she was to say to Marianne. There were ladies staying at the hotel armed with walking sticks and shod in stout shoes who did not seem to find it strange that one might wish to walk. She followed the path they had pointed out and wished Guy could see her as she tripped along in her grey suit, her expensive leather handbag held loosely on her arm. So much better than the old-fashioned reticule which she had always scorned, preferring to be free of incumbrances.

She followed the winding path that led up and then down, skirting the contours of the lake, and passed over a little stone bridge. She paused for a moment to look over to where a tiny stream lapped over stones and moss. The line of trees in the distance was, she had been told, on land belonging to Langthwaite and if she went through a little coppice she would come out on another lane that led to the orchards belonging to the Mortimers. What would she have felt, what sort of life would she have had up here far away from worldly pleasures if she had married someone like Ned? Before entering the wood she looked down and across the lake to the wooded islands that floated in the still grey-brown water. All was quiet and peaceful. She was glad that she had come in summer, for winter up here would be scarcely bearable. The air was fresh and sweet-smelling and a bee passed her, busy on its own mission. It would be hard to make a garden grow here, she thought, in this wild rugged country. Even so, she would admit it had its own beauty, though not one she would wish to live among, preferring the warmth and rich verdure of pastures leading down from other mountains to villas with orange blossom under snowy peaks, larger skies. Everything in England was on a small scale.

Miles liked it. At the thought of her son she stood for a moment leaning against a stone wall which was hard against her back before walking briskly through the coppice and through to Langthwaite Lane. She had a picture of Miles in her mind. Miles fishing, riding, boating, swimming, enjoying a country life, perhaps becoming a country doctor. That sort of life would not bore *him*.

The gateway to the old-fashioned Georgian house that was 'High Pines, Langthwaite', was open, and Kate went through it and on to the short drive that led round to the front. There she paused a moment, then took a deep breath and pulled on the bell. There was at first no sound except for the bark of a dog far away and the rushing sound of the beck that fed the lake and descended somewhere in the grounds of the house. She could just see the lake sparkling beneath and a small sailing dinghy with a blue sail rounding the point opposite the island. Then the door was opened by a person she recognised as the 'Polly' who had once visited Lake Hall Park with Marianne and her family. Now that Kate was on another's territory she knew she must keep her head. She was shown into a small sitting-room where a firescreen in front of a white fireplace proclaimed that for once the weather was warm enough to dispense with a fire. On the table in the window alcove was a vase of red and white flowers and there was a long, glassed-in bookcase that bore witness to the polishing habits of North-country servants. She was looking at a watercolour sketch of the lake when Marianne came in at the door. Kate turned and was greeted. Marianne had the impression that it was she who was visiting perhaps a doctor or a dentist, for she felt awkward in her own dear familiar room. Kate saw in one glance how pale and thin she was. They shook hands warily and Kate sat in a comfortable armchair while Marianne's slow steps reached a small sofa and she sat on it with every appearance of thankfulness.

'I was sorry to hear about Ned,' Kate began.

'It was a shock,' said Marianne faintly.

'I hope you are beginning to recover from it nevertheless?'

'I have not been well,' said Marianne, but trying to give a polite smile at the same time. She dared not ask her why she had come.

Kate would have answered her own unspoken question for her, but the parlour-maid appeared wheeling a trolley with a silver teapot and plates piled with buttered scones. Kate took one and bit into it heartily and suffered a cup of tea to be placed near her on a small three-legged table by the chair. How different from her interview with Sarah, where she had felt just as uneasy, but in her own ambience.

They both looked into each other's face searchingly at the same time, but Marianne still said nothing. She had not taken anything to eat for herself and had placed her hand by the silver teapot as though she needed its warmth.

Kate decided after a taste of the tea, which was strong and pleasantly aromatic, that she would be frank.

'I am just as nervous as you are,' she said, looking at Marianne meditatively.

'Did you come about Louisa?' Marianne's voice choked on the words. She felt as though she were about to have a heart attack or that her lungs would not take in enough air so that her words came out in a gasp.

'Partly – it's my son's happiness I am most concerned with, as I am sure *you* are most concerned with your daughter's.'

'Yes,' whispered Marianne. She made an effort and looked Kate squarely in the face. 'I would not allow her to become fond of him – you must know the reason.'

'That she is Guy's? Yes, I know.'

'How long have you known?'

'Sarah said nothing to you then? About the fact that I had guessed?'

'Sarah? No. Have you seen her recently? It was Sarah – I asked Sarah to tell him. Louisa knows nothing.'

Kate let this pass. She was not yet ready to tell her friend that her son was at present at Lake Hall Park.

'Yes, Sarah told me.'

'Then you must agree? Don't torture me, Kate – I can't alter the past. For more than twenty years I have been trying to forget. When Ned died it all came back.'

'Then Ned never knew?'

'Oh, no! No! I could not ruin his happiness – and mine depended on him.'

Kate wanted to get up and walk around, but composed herself and tried to think how to continue. She decided to think aloud for a moment and let Marianne try to understand.

'You see, Marianne, it has taken a lot of courage – I think they call it moral courage – for me to come here to you. I should have come years ago, but I had not the courage then, as you had not when you married Ned. What I am going to tell you I can only tell you if you will promise that no one but Miles and Louisa and Sarah will ever know. Above all my husband must not know. You will see why when I say that – ' She paused for a moment and Marianne regarded her earnestly. Now that it was out she felt a little better. Kate would know what should be done.

Kate changed tack. 'I intend soon, when my daughter Kitty is married, to return to the States with Guy. I think I have had enough of Europe – perhaps the old pioneering blood will be out. Miles will probably stay in Europe,' she added carelessly. 'He loves England, you know – '

'But Louisa! That will only make it worse for her – not that I believed she was in love with him, and anyway that is all past. I don't talk about her feelings to her. We've grown apart since Ned died and I forbade her to see your son. I was frightened, you see.'

'Marianne, your daughter *was* loved by Miles, I can assure you of that.'

'Oh!' Marianne's hand flew to her throat.

Kate chose a roundabout way to enlighten her further. 'If it is of some comfort to you,' she said drily in the old Kate manner, 'It is not Louisa you have to worry about.'

'You are torturing me, Kate. As though it is not enough to know I sinned, that Louisa did not belong to my dear, dear Ned, and now you say that her own half brother "loved" her! What is to become of us all?'

Kate went on. 'It is your other daughter – Emma is her name isn't it? – *she's* the one you would have to worry about so far as Miles is concerned. But as he's never met her but once, it doesn't signify.'

'I have to worry about Emma! But why? Don't talk in riddles, Kate.'

Kate finally got up and walked to the table and touched the flowers and then turned once more to Marianne who was now sitting up on the sofa, her eyes large and imploring. 'Why, you goose, *think*! Because they had the same father!'

It was Marianne's turn to get up. 'What do you mean? What are you saying?' Her friend's words could not seem to sink in.

Kate's face was white in spite of her bantering tone. 'This is the bit that needs "moral courage",' she said grimly.

Marianne waited, staring at her. Kate came up to her again. 'Sit down. I will sit beside you.' She cleared her throat. 'Miles is not Guy's son, Marianne.' She said it slowly and clearly, but Marianne still looked dazed. Her words echoed round and round the room and then Marianne's hand was taken by Kate.

'Ask me then. Ask me whose son my Miles is, *ask* me.'

'Are you speaking the truth, Kate?' asked Marianne feebly.

'Of course I am. I should have enlightened you years ago. But, like you, I could not.'

'But why, why *Emma* then?' Kate was silent. 'You mean that Miles is – is *Ned's* son, Kate?' whispered Marianne.

'Of course. He is your little Emma's half-brother – but as I told you, he has no designs on her.'

Marianne stared at her and went on staring, unable to credit her ears. She groped as in a mist. 'Emma is Ned's. Miles is *Ned's*?'

'I see I shall have to go into unpleasant details,' said Kate. 'Give me another cup of your tea.'

Marianne almost overturned the teapot and the cup she was proffered rattled in its plate.

'I know Louisa is Guy's daughter. But Miles *is* Ned's son. They are not related,' said Kate in a quiet voice when she had drunk from the cup and put it down again. 'We shall have to forgive each other,' Kate went on hastily, having ascertained that Marianne was taking in her words.

Marianne felt as though a great weight which she had been carrying round for a quarter of a century had begun to roll away. But she could not let it go quite yet. She gasped slowly as though solving a difficult problem of logic. 'Then you – and Ned . . . ?'

'Don't say you always thought he didn't like me,' said Kate. 'He *didn't* like me – you would be quite correct. But that need not worry you.'

'Did he know?' asked Marianne, seeing the whole of her married life in a different light.

'No, like Guy, he didn't think. Men have this in common: when they have done something they are ashamed of, they forget it. Not like women.'

'Louisa said that Miles was like an Englishman, that he liked the countryside . . . that was why you sent him to Ned's funeral!'

'I thought he ought to be there – yes. He had no idea why, of course. Last week when he returned from Boston I asked him about his feelings for Louisa. I wouldn't have told him anything, anything at all, if he hadn't told me that he loved her, that he could not bear to have lost her, that he had been told by Sarah at your instigation that she was his sister. So I had to choose – '

Marianne was still thinking about Ned and Miles, remembering that Louisa had said he was 'like Papa – interested in technical things'. Somehow this little detail made her realise that it was indeed true. But how? How could Ned and Kate . . . ?

'I didn't *have* to tell you,' said Kate. 'Like you I thought the little youthful passion would blow over. But they are not children. Miles is a young man, Louisa a young woman – older than you and I were when we – ' She stopped.

Marianne brought her thoughts round to the woman who was sitting before her. Kate Mesure. Kate Demaine. How she must have humbled herself to tell her this. Yet she could have spared her all those years of misery with just a whispered word. 'Was it because you loved him, Kate?' she asked timidly. Her Ned. Ned, who had saved her reason and her honour. And had never known. But how could he – ?

'I suppose I'd better tell you about it,' said Kate. 'Or you'll be thinking I've been pining for him all these years – Ned, I mean . . .'

Suddenly Marianne realised that one part of her did not want it to be true. One part of her clung on to the idea of an irreproachable Ned, the Ned whom Sarah too had loved, Ned who had instead proposed to *her* and loved her and his family. Ned lying out there in the graveyard.

'But why?' she cried. 'And why if – if he went with you, why did he come to me – and why pretend he didn't like you?'

'I told you – he did not like me at all. I suppose – I *know* it was my own fault. Just as it always is the woman's fault, isn't it?'

'He never breathed a word to me,' said Marianne in a small voice.

'But neither did you.'

'So our marriage was, ultimately a deception?' She lifted a shocked face to Kate.

'Marianne – we don't live in an ideal world. I'm sure your marriage was as happy and successful as mine is. That is, not perfect, but we suit each other.'

'You must tell me how it happened,' cried Marianne. 'I have no right to know but I must know. You can imagine how Guy seduced *me* – my own fault, as you say – but you were so much more sophisticated than me, so much more in charge of yourself – '

'I wanted to tempt him,' said Kate simply. 'And I got my deserts. I was "worsted in combat". Served me right. So I had to marry Guy, you see, rather quickly. But I swear I had no idea he'd seduced you. I may be immoral, but if I'd thought he wanted you, or that you were to have his child, I'd never have acted as I did. Anyway, you were in London and I was on my way to Baden with my uncle and aunt and I got them to invite Guy to dinner in Paris. He'd just got back from England. It was quite simple. I suppose both our babies arrived rather early?' she added with a slight smile.

'But did you not love Ned, Kate? He was honourable. He'd have married you.'

'No, I did not "love" him, and, yes, he would have married me before we – and a miserable life I'd have led him. Instead, he had all this.' She gestured round the room and out of the window. 'He was attracted to me, of course, but I think he despised me really. And I made him despise himself and that is what men can't bear. We can say things here which we need never say again, Marianne. I came to tell you that Louisa and Miles have my blessing. But I find I want to confess to you. Strange, isn't it?'

Marianne was trembling and her mind seemed to be whirling again. Babies, they could have exchanged them, but of course babies belonged to their mothers more than their fathers and after all Miles was Kate's and Louisa was hers . . . The men had only been enablers, progenitors. She and Kate had borne their children, nurtured them in families whose most sacred code, that men should know their own children had been broken. 'It's a wise child,' she cried hysterically. Then she recalled herself. 'I'm sorry – all this is so unbelievable.' How had the cool Kate been reduced to the same expedient as herself? 'I did love Guy,' she said instead. 'I wanted him to love me. I was foolish, wrong, I thought I didn't live in the world of marriages and money and social obligation, though of course I wanted him to ask me to marry him. I thought we were having just a time out of time, that he would go on loving me, or that it would be the happiest memory of my life, even if I lost him. I don't know how I managed afterwards. It was such a surprise when Ned sought me out. I couldn't believe it – still don't quite understand. Though we were happy, Kate, we *were* . . .'

'Of course you were. Ned was the right man for you,' said Kate. 'And Guy for me. In fact it was my plan to get Guy in any case. I thought I'd make him just a little jealous if I flirted with Ned who did find me wicked but attractive if you must know. The silly thing was, I could have had Guy in any case. I'm sorry to speak bluntly, but he would only have married a girl with a fortune.'

'I still don't understand about Ned. I *want* to believe it – it would be the best thing in the world for Louisa, if, as you say, they do love each other. But I'm terrified she'll make the same mistake I did.'

'Miles is like his real father in many ways,' said Kate. 'But he has not been brought up to think that sexual passion is something shameful as Ned was – as all Englishmen are. It was only the once, Marianne, as I believe it was with you and Guy.' She was silent for a moment, remembering. Then she lifted her head. 'We need have no further secrets. Do you remember the day you conceived Louisa – Sarah told me – it must have been that very afternoon. Don't you remember, Ned and I going out riding together? That's when it happened. I taunted him, got his blood up. I did not understand as much as I thought I did about the way men are. I tempted him, made him angry, then made my horse gallop away, but he followed me and when the horse threw me – not heavily – I was very reckless. I suppose he found me irresistible. I'd only wanted to tame him, make him mad for me and then amuse myself treating him badly. He was such a prig, you know. I'm sorry, but he was. He made *me* angry too and I found I had a sort of power over him which truly I didn't want. I'm not a sensual person. Not like you, Marianne, perhaps that's why I took the risk. I wanted to see how far he'd go.' She was silent again, remembering how Ned Mortimer had been so angry and ashamed afterwards and how she had had the last word. 'I still remember what I said to him. He was too good for me, you see, but he felt I reduced him to the level he thought was mine. I said, after it was over, and we were both panting, and he was weeping – he had *fought* me, was mad for me – I said "Don't ever come near me again or breathe a word of this to anyone. I am soon to be married. Now go." I waited till he'd remounted and ridden away and then I went straight in to Guy, who seemed a little languid and uninterested, and I thought "I must have this one. *He* isn't ashamed of lust. *He* isn't filled with notions of honour. He isn't an English brute when the surface cracks.' We were two of a kind, Guy and I – yet I half feared, half admired Ned, as well as hating him for what he had done. I'd been quite near it before, of course – in Italy – but he was my first man. How I fought him and yet knew that my fight was another way of tempting him. I knew that even as he overcame me. I wanted to see him lose his gentlemanliness, his inhibitions you see – perhaps I shouldn't have told you all this . . .'

But Marianne was remembering the Ned she had known. If only they could both have confessed. Now Ned was her posthumous equal in deception and the burden that she had been carrying was lightening upon her shoulders. But it had been succeeded by another burden, one that she knew no confession could ever remove. One she would now have to live with. If Ned had known about Guy he would never have married her, she was convinced. She, though, could perfectly well have married him. Men 'sowed wild oats', women 'fell'. She and Kate had done the only thing possible. But why had Kate not wanted Ned?

'I told you,' said Kate, voicing the answer to Marianne's unspoken question. 'I should have been the wrong wife for him but, even worse, he would have been the wrong man for me. I'd have ruined him and he'd have put up with it because he was a gentleman. I only got what I deserved – I don't suppose it even occurred to him I might be pregnant as a result. He married you so soon afterwards, didn't he? I'd have made Ned so miserable – give me a little credit, my dear. Guy, of course, I'm sorry to say, was only too delighted to find I wanted to marry him.'

Marianne shivered. How could two intelligent women each have managed to get herself seduced by the wrong man?

'Lust is more powerful than love,' said Kate.

'You think that's what Ned's son feels for my daughter?' said Marianne with a spark of indignation.

'No. Miles is enough of a romantic and has enough of his father in him to want to like the woman he desires. Miles would never act like an angry bull.'

'Let us walk in the garden,' said Marianne impulsively. Both women rose and Kate followed Marianne out of the room through a little door at the end of the passage which gave out on to a lawn and trees and overgrown rosebeds. Ned's roses, thought Kate. It was quiet and peaceful. 'You needn't have come,' said Marianne out of another silence when they were both disposed upon a wrought-iron garden seat.

'No, I nearly did not. Perhaps I was frightened that Guy would one day discover about Louisa, and even now he must not know. If he did, he would have to oppose our children's love, unless he also knew of my own past and that Miles was not his son. I don't mind having told you, but I cannot have Guy know. Then I should have no hold upon him. I thought, I can only stop your little girl's misery – and Miles's – at the price of my own reputation in my son's eyes.'

'Think what it will mean to him to discover who his own father really was!' said Marianne. 'How will he deal with that?'

'At present he is awaiting a telegram from me to say he can go to Lake Hall Park where I believe your Louisa is,' said Kate complacently. She had better not tell Marianne it had been her idea for Sarah to invite her.

Marianne caught her breath. 'Was that wise?'

'I have also written a letter to my son explaining everything. I shall post it to him in an hour with your permission, to await him at Sarah's.'

So even Kate had shirked seeing her son and preferred to administer the shock at safe distance, thought Marianne.

'With your permission he will tell Louisa the whole story,' said Kate. 'Unless you prefer to go to see her – or write to her yourself?'

Marianne was thinking how Kate had worked it all out and put her plans into action.

'I know my son better than I ever knew his father and I am absolutely certain that once he knows the whole truth he will know exactly what he must do. I envy him – and you too, Marianne – for believing in "love".'

'You think they will wish to marry?'

'I don't know – at least it will be without illusion on either side.'

'Louisa stopped talking about him. Oh, I have been so blind. What shall I do now?'

'I would write to her, just to say that Miles will explain it all. Once she knows, you will find it all easier to deal with. It is up to you now, Marianne, as far as your relations with your daughter are concerned. That is not my business.'

'No, it is hers, I suppose. Will Miles be *angry* with you, Kate?' she asked timidly.

'Louisa will not be angry with you, Marianne. She has seen how you have suffered, I expect,' replied Kate, not answering the question. For it was true she did fear a little what her son would think of her. Even kind, tolerant Miles could hardly be pleased he had both gained and lost a father.

For a moment they sat there in that remote northern garden which Marianne had neglected for the past year. She saw their children's lives like some complicated game of billiards with different coloured balls, knocking each other into the wrong pockets, but miraculously ending up in the right ones.

'But life is not a game,' she said aloud.

'It has often been treated as one.' Kate changed tack. 'Louisa and Miles, according to our old friend Sarah, actually do like each other. Is not that even more important than "love" if they are to survive marriage? You liked Ned. I even like my incorrigible Guy.'

They were silent, each thinking complicated thoughts that were better left unsaid.

'We need not communicate again till things are settled,' said Kate as she left, having refused to stay any longer than was necessary. 'I expect you will need more than a few days to accustom yourself to the change in your state of mind.'

She had returned to being the cool-as-cucumber Kate and afterwards Marianne was hard put not to see the whole afternoon as a dream. Kate when she was once more alone wondered if she had uncharacteristically said too much. But she had done the right thing. Of that she was absolutely sure. She had released Marianne from guilt and at the same time sacrificed her son's idea of her and of his father. Well, Miles was grown up now and a man. He had better act like one. But she thought too that if the world had been different she might not have needed to marry at all. She would give some of her investments to Miles and see whether he still preferred to work for his living!

She sent off her telegram to Paris to Miles and posted her letter to Lake Hall Park to await his arrival. Then she took the night train back to London and did not sleep, wondering whether she had told the whole truth of her feelings. It was true she had wanted that afternoon, so long ago, to 'tame' Ned Mortimer and instead had been humiliated, whilst he had been, he thought, 'corrupted'. Let Marianne think over her past married life and perhaps she might find that having to forgive Ned's lapse would allow her to forgive herself for a similar one. Had she herself lost face? She would dearly like to have known Marianne's secret thoughts. She had made up for the past in one stroke. Now the present was to be confronted and tomorrow she would turn her mind to it.

Marianne's thoughts were in fact different. Once the burden of guilt had fallen off her shoulders, she began to see Ned in a rather strange new light and to try to remember incidents from their early married life that might have given her a clue. How guilty poor Ned must have been, how full of compunction. At least she knew she had made him happy. There was only Louisa now to face. The others need never know the truth. Guy she dismissed from her mind as Kate's responsibility. She wrote a short letter to her daughter, but could not keep her hand from trembling.

> Dearest Lu,
>
> Everything is going to be all right. By now Miles Demaine will be with you and can explain everything. His mother came to see me today and has written to him. You must forgive me, darling Louisa. We will talk of it soon, but not now. Be happy.
>
> Your loving
>
> Mama

Marianne crept out alone the next evening and walked over to the churchyard with a bunch of roses and sat long by Ned's grave. Alex Carmichael came upon her there, having gone up to the house and been told where she had gone. He was pleased; at last she seemed to be mourning his old friend in an ordinary way.

23

Kate Demaine had other things on her mind when she arrived back in London. Emotionally exhausted as she was, there was another shock to come. A telegram from Guy, who had returned to Paris, filled her with the utmost foreboding. She would have preferred it to be a demand for money for his gambling debts and it was small consolation to know that it was not.

'Come immediately. Kitty eloped.' was the cryptic message.

Who on earth had Kitty eloped with? – and what had Guy been doing for such a thing to have happened? Kate arrived in Paris in a state of anger and dismay which effectively precluded her thinking over in too great detail her long confession to Marianne. Perhaps her own lies were coming home to roost at last?

On arrival at the apartment she found Guy actually trembling with what she recognised as a certain well-merited terror of herself. What had happened was that Miss Kitty had made her own assignations and plans when Guy had been at the gambling table. She had been charmed and captivated by the liquid brown eyes and even more liquid voice of the tenor of the small opera which played out of season on the Côte. One night Guy had arrived back at their hotel to find Kitty gone and a scribbled message in his room. On enquiring from the hotel staff: 'Yes, they had seen Mademoiselle Demaine with her friend Signor Frescobaldi and Mademoiselle had seemed in a hurry to get off to the station'. They thought she was also with another lady, an older lady, who was chaperoning her. Who the devil could that have been, Guy wondered, but then realised that the couple must have laid their plans well in advance and taken along another member of the company. The message had said only that by the time he received it she would be in Naples and please to inform Mama. The couple would return to Paris married and hoped they would be forgiven.

'The devil they will,' said Kate. 'I expect he knows of her fortune. Didn't you even guess what she was doing?'

Guy, penitent and distressed, was all for getting the police on their tail, but Kate, thinking it over, decided they would await events. What price the Prince's son now? The whole thing was preposterous. 'I can't send Miles after her,' she mused, when Guy, by turn blustering and enraged then tearful, had actually taken her hand promising he would never again let Kitty out of his sight once she had returned, that he would have the ruffian thrown into prison and that he had not gone after them because he thought Kate would know better what to do.

'She's only just seventeen,' Kate said coldly. 'Yet her reputation may not yet be quite in shreds. What matters is whether she has actually got herself married, don't you see? I ought never to have left her with you. You have no idea what young girls get up to. We shall go to Naples and see her and settle matters there. She will have to have visited the French Consulate to get permission to marry as she's under age. I doubt they will have given it. She will force our hand. Don't you see that?'

Guy, roaming round the apartment in his *robe de chambre* could only protest feebly that she was only a baby, that the man should be hung, drawn and quartered.

'Too late,' said Kate. 'If she insists on marriage we shall have to see what this Frescobaldi person is like. Do you know anything about him? How old is he?'

'Oh, quite young,' said Guy. 'At least that's what they said. *Quelle horreur*!'

Off they went to Naples the next morning and Kate could not help reflecting how different this journey was from her voyage to England. But she also reflected that Kitty had perhaps done what she herself might so easily have done and what Guy had often done with no thought of marriage. Was she to be hauled back disgraced or allowed to marry?

The Consul was helpful. Yes, Mademoiselle and her fiancé had been there three days before. Papers were being prepared, but Mademoiselle had been told that she needed parental permission. Mademoiselle had cried and the gentleman had comforted her. He was a fine young man and talented too. Ah, young love! He hoped that matters could be settled to please all parties. Kitty's address was located and the irate parents drew up before a peeling *pensione*, but in a respectable quarter, only to find that the young couple were out. 'We shall wait,' said Kate. Towards evening a barouche arrived and from the caretaker's room they saw their young and lively daughter helped out of a carriage by a handsome young man who looked, Kate thought, a little like Guy.

Three hours later, after tears and defiance on one side and a fine display of temper from Guy and a coolly appraising silence from Kate, Kate had made up her mind. Franco, for such was his name, protested undying love for their errant daughter and a touching desire to put his operatic earnings at her disposal, but Kate was exercised only to know whether Kitty was in fact technically, whatever might be the socially conventional case, 'ruined'. She requested a private interview with her daughter which took place in a room put at their disposal by the manager of the *pensione*.

'I don't care what you live on,' she began. 'And I have nothing against Signor Frescobaldi, though I thought I could do better for you. But I must know whether you have yet succumbed to his charms.'

'Franco wanted to wait until we were married,' gulped Kitty. 'He is a Catholic.'

'So, the elopement was your idea, was it?' asked Kate conversationally.

Kitty did not answer at first. Then, 'You would never have allowed me to see him. You wanted me to charm that horrid Prince Vrigne's son. I hate him. If you make me come back with you I shall never speak to you again – and I'll never marry *anyone*.'

They had never had a conversation in which either showed her feelings since Kitty had been a toddler. Kate seemed to be giving the matter deep consideration. Then she said, quite kindly, 'Perhaps it was a mistake to want you to marry well. I expect you "love" him?'

'Oh, yes, Mama!'

'Right. You return home for three months. You may see your fiancé during that time in Paris – all above board. If he still wants you and you him we announce the wedding and you marry from home. No harm has been done. You might change your mind. If you don't, well it's your life. You're too young, of course, and I'd rather you didn't marry yet. But I promise you will receive our blessing if you will wait. I'm sure *he* would rather do that.'

'Franco wanted to write to you before,' said Kitty virtuously.

'But I told him you and Papa were snobs and would disinherit me. He said he didn't want my money. He wants me.'

'Then you will do as I say? Let us hope the news hasn't spread.'

After a silence Kitty said, 'Yes,' quietly, and then flung her arms round Kate. The matter was concluded, no one, apart from the family any wiser, and Kitty returned to Paris with her parents, sure of her swain's undying love.

'Of course he will still want her,' said Kate to Guy. 'Now he gets her *and* a settlement.'

Guy had been astonished at the turn things had taken.

'It's sometimes better to let young people have their heads,' was all Kate said.

24

While all this was happening on the Continent, much was afoot at Lake Hall Park. Miles had not yet arrived when a letter came for him in what Sarah recognised as Kate's hand. Louisa saw it and plucked up her courage to ask Sarah why anyone should think he was at the Hall. Sarah, who chafed continually under the knowledge that it was she who had been instrumental in removing Miles from Louisa's life, and she who had promised Kate to invite both of them to the Park, none of which she was at liberty to disclose, mumbled something about Miles perhaps coming to stay for a few days, which made Louisa suddenly pale.

'You mean he has asked to come?' she enquired of Sarah later, under cover of gathering some strawberries in the kitchen garden with her hostess. Olivia was busy with her prep.

'I expect it will all be explained here,' replied Sarah, diving into her capacious apron pocket and producing a letter for Louisa in Marianne's inimitable curly hand.

'From Mama? Oh, I do hope she is not worse. I am so worried about her.'

'Go in and read it, dear. I don't know the whole story, but I have to confess that it was Kate's idea you visited us this summer. I believe she is going to – to – send Miles here.'

'Kate! But . . . ?' Louisa was just about to say that it was all Kate's doing that she and Miles had been parted, but she held her tongue. She took the letter into the tack room where Sarah kept a desk and an armchair. Somehow she did not want to go into the house and she was impatient for her mother's news. She had not written her a proper letter for ages and when they were together they hardly communicated. She opened it in fear and trembling. One glance through and Louisa sat down rather heavily on the old leather chair. One of Sarah's dogs strolled in and whined, but she did not hear him. Then she began to read it again attentively, incredulously. Miles was to come! But how could everything be all right? Had her mother known all along something she did not? The more she read it the more puzzled she became. Would Miles really come? She could not wait, but ran out of the tack room to find Sarah.

'He should be here this afternoon!' Sarah cried to her. 'I've just got a wire from Dover. Are you very fond of him dear?' she asked Louisa shyly after a pause.

For answer Louisa held out Marianne's letter and Sarah saw that her eyes were brimming with unshed tears. Sarah put on her goldrimmed spectacles which accompanied her everywhere, read the letter and handed it back to Louisa. What on earth had happened to change things? Kate was at the bottom of this.

Louisa was thinking along the same lines. 'Why should Aunt Kate visit Mama?'

'I don't know, dear – except it would not be for something bad or your mother would not have written to you like this.' She was all at sea. Nobody had written to *her*.

'Why did Miles stop loving me, Aunt Sarah? Do you know?'

Sarah was dumb. 'I can't say anything till Miles arrives,' she finally got out, and fled the garden. Louisa looked at her in astonishment. Had everyone gone mad?

She ate little lunch and conversed only with Olivia, who seemed oblivious to any undercurrents of excitement or anguish across the table. As soon as she could Louisa escaped and walked up and down in the rose garden. When would Miles arrive? Oh, let him come soon! Even if he doesn't love me . . . She could not seem to calm herself. Yet Mama had said things were going to be all right. For a moment she wished she were Olivia who had no secret love (as far as she knew) and who was at present peacefully indoors writing her English composition. How was she to greet him? She felt really frightened as well as excited and decided to walk into parts of the garden where she had never been before. She must have a little time to herself. Let him find her if he has anything to say to her.

She was sitting on a rustic bench later that afternoon when Miles Demaine came looking for her. She was not in the rose garden but in a part of the garden that was hidden from view of the Hall where a small lawn had been laid near a little wall that bounded the property from the park proper. It was a hot afternoon and he was still a little dizzy from his journey, had come straight on from London after a tedious night crossing. A sandy path led to the lawn by overgrown Bourbon roses, making a hedge on each side. Trees bowered the garden here on the park side and their shade was delicious, cool and green. The lawn was set in what might have been a forest glade and in the middle of the lawn there was a small stream that had wandered through from the park. A plank lay over it in the middle and there was a sundial by it and under the far trees a seat on which Louisa Mortimer sat, the leaves dappling her face and dress. She was as still as a statue as he approached. Miles knew he must play his cards carefully, but was so agitated that all he could think was that Louisa was alive, was here, and could be talked to. He came up to the seat and she looked at him with a tremulous smile.

He had never known his mother tell an outright lie. Self-serving she might be, but always honest with others, at least if one discounted sins of omission. He was holding her letter to him in his hand, the letter which he had just opened and read with increasing disbelief, then with relief. What was he going to say to Louisa? How did one begin such a conversation? He should have gone away somewhere quiet first to think it over, but he had come straight out looking for her. Would she even want to see him? Did she know *nothing*? When he came towards her, and she looked at him with that smile of hers, he hesitated a moment. Then he sat down beside her. 'We must talk,' he said. Rapidly and with face averted he told her everything, not daring to look at her till his recital was over. She sat like a statue; dismay, incredulity, then joy rapidly following each other across her face, but when he finally stopped speaking she turned his face to hers with a small hand and said, 'It won't make any difference to us, all that. Not now, will it?'

He felt a sudden rush of relief and all he could find to say was, 'I love you, Louisa. You will marry me, Louisa, won't you? I mean, don't decide now – your head must be spinning – I know mine is. But soon – this year?'

'I love you, Miles,' she said softly. 'Big, almost brother, I love you and, of course, I shall marry you!'

Then he took her in his arms and for a long moment he neither kissed her nor said anything at all, but just held her enfolded. She too clutched at him, would not let him go and she was crying. Gently he released her and sat down on the little bench holding her hand.

'Don't say anything else. Just be there with me. Don't go away. Oh, Miles.'

It did not seem to matter that their parents' lives had been carried on in such a potentially tragic muddle. He took her other hand and then brought both her palms to his mouth and brushed them with his lips. They remained for a long time silently, just looking in each other's eyes. 'I *hurt* you. It was terrible when I went away,' he said and hid his head on her shoulder.

She took his hand to her lips.

'I don't mind about my Papa – I mean about my mother's husband. Not really. It seems a dreadful thing to say, but I've never felt I loved him,' he said.

'I loved *my* Papa,' whispered Louisa. 'Papa loved me too. I'm glad he didn't know I wasn't his. But he never knew about you either. Our poor Mamas!'

'If I could have chosen as a child I would have wanted you to be my real sister,' he said softly. 'But I *am* glad you are not.'

The hot afternoon, the scents of summer seemed to encase them both in a shimmering glass globe and never had Miles felt less like a brother. They looked in each other's eyes and then they kissed each other softly again.

'I thought it might take a bit of getting used to,' murmured Miles into her hair. 'But you seem all right. Just the same. Just Louisa. You haven't changed.' I expect the shock may come later, he thought, and she will be tearful and angry – not with me, but with her mother and all the deception.

'Are you angry with your mother?' asked Louisa as though guessing the direction of his thoughts.

'Why? It is no one's fault. Not even your – my – father's.'

'Loving is not wrong. But if only Mama had told me the truth . . .'

'You would have been even more unhappy. Like this one thing cancels out the other . . .'

'How sad you must have been, Miles, when you knew only one half of the story. Does Aunt Sarah know?'

'Only the part about your mother and my – Guy.'

'Then we shall have to tell her the rest or she will think herself obliged to stop our meeting!'

'I expect they will write to her. She trusts me, I think.'

'That must be why people think I am like Mama,' said Louisa, after a pause. 'They are always saying so. Yet I don't suppose anyone guessed . . .'

'In what way like your Mama?'

'That I give way to feelings. Mama must have loved Guy, you know. She is very romantic. Yet I don't feel I want to meet him,' she said. 'If I hadn't met you we might never have known, might we? They would have kept it all secret.'

'Not even Sarah knew till last year.'

'Till my Mama told her?'

'I wonder if anyone else did. I don't suppose so.'

Louisa was thinking that people did not really live in the world of convention and politeness and marriage and 'things', but in the world of feeling and memory and dreams. How terrible when these two worlds clashed. Why could people not be honest about love? Was it so awful a thing that it must be ignored by polite society except as a matter of titters and deceptions? She knew suddenly that her mother had felt the same once. Her mother had loved Guy Demaine. She must have done. How dreadful, then, to have to marry someone else. Had Mama loved Papa too? *Their* life had been about things and conventions and ordinary events and obligations and they had seemed happily married . . . ? She could not imagine her mother as a young woman, younger even than she herself was now, getting herself in an awful fix and being rescued by Ned Mortimer. Perhaps people had reason to be frightened of passion. She remembered the eyes of Aunt Sarah's husband, looking at her the way Monsieur Demaine must have looked at her Mama, and shivered. And she, Louisa, was the result. How amazing. Mama had 'fallen'. But men did not 'fall', did they? She found she was firmly on the side of her Mama, in spite of everything.

'Let me look at you again,' said Miles, breaking into her thoughts. 'You are so beautiful, Louisa. Do you really love me?'

For answer she kissed him again. 'Loving is quite easy,' she said. 'And I love you. Let us go and find Olivia,' she said to break the tension. 'I have to give her an English lesson – that is the reason I was asked to come.' She got up and he gave her the book which was lying unread on the bench.

'Swinburne,' he murmured.

'Time with a gift of tears,' said Louisa. 'But for Mama – not for me.'

They walked across the stream and Louisa stopped to smell the roses.

'I shall have to talk to Aunt Sarah,' said Miles, but knowing there was so much more to say. It was like a dream, but usually a dream did not have to be tested out on others. This dream was true and something they had to assimilate together. 'It won't spoil our love, will it – all the past?' asked Miles again stopping suddenly as they walked by the pool.

Louisa thought before she answered. 'I think it will make us even more grateful – or at least me – to have someone who knows all about me! My Papa and your Mama – it's like a fairy story or a dream, isn't it? But I don't want to live in a dream, Miles, I want to live with you.'

'I'll wait for you,' he said. 'It may be some time for you to sort out your feelings . . .'

'No need to wait for me,' said Louisa and put her arms round him again. At that he began to kiss her fiercely and she returned his kisses. Then they drew apart and walked along towards the house. By the end of that afternoon the whole world was a different place for both of them.

For the first time for almost twenty-five years, Marianne felt the past, like Ned, had really slipped away, and with it all the feelings of her youth, except for a slight melancholy aroused by the memory of them. She remembered too, as though it were also long since, lying on her sofa feeling ill and weak, refusing to confront her own actions; remembered too the nightmares that had pursued her sleep. Resignation was taking the place of anguish, along with the relief that now, now at last, she might become the self that had eluded her for so long. She was nearly forty-two, old enough now to grow up.

She went daily to Ned's grave, sometimes with flowers, sometimes empty-handed, and sat there by the mound that covered him, not always thinking of him, but putting together her own life. She even planned to have the old mare saddled and to ride along the lower paths by the lake or in the copses as she was no longer afflicted with that fear of leaving the house that for the year following Ned's death had cramped her spirit. She thought about the ideas that had galvanised her youth, and planned to settle down when the autumn came with some reading. She could even begin to envisage a future for herself! Yet she had been punished, and still was. To the end of her life her happy marriage would have the double shadow of Ned's and her premarital conduct over it. She felt Ned's shame and guilt in herself and bore it for both of them. All that mattered now was Louisa's happiness.

Louisa and Miles came into tea and then disappeared.

'Where are Miles and Louisa?' asked Sarah, who had not come down to tea and was waiting to see what happened.

'They left a note,' said Olivia placidly. 'Gone off to Oxford, I think – they seemed awfully excited.'

'Took the mail phaeton,' said Thomson disapprovingly when asked. 'Saw them go like they were in a dream.'

Sarah opened the note, signed by them both. 'Darling Aunt Sarah – Everything has been explained. We are not related. We have to be alone for a time. Don't worry. We'll return by the end of the week. Louisa and Miles.' Scarcely had Sarah time to digest this when Thomson came rushing in again.

'It's the Master – come on a visit. Oh, Madam, I do wish you'd told me.'

And there was René getting down from a bay mare which he seemed to have ridden from the station for she recognised it as the rector's, who sometimes lent it out. 'I was bored,' he said. 'Thought I'd pay you a visit.'

'I didn't even know you were in England,' she said stiffly.

'Have to see my wife now and then,' he said lightly.

'Hello, Papa,' said Olivia. 'You've just missed Miles and Louisa.'

'Really? Then you will be able to concentrate upon entertaining *me*, I suppose,' he said. 'What about a ride tomorrow? The rector has another mare – a nice grey – unless your mother's got a better horse for you.' He was always awkward with his daughter.

Olivia sighed. 'Will you take me to London with you next time?' she asked. 'I want to sketch the costumes they have in the new museum.'

'Will you see Mama?' Sarah said. 'She will be so pleased you have come.'

'I doubt that, but I'll do my duty,' he replied. He seemed in a good mood now. She wondered whom he had been seeing in London. 'Came to talk business with your Mama,' he said to Olivia and Sarah's heart gave a small jump. She had intended to wait until her own mother took a turn for the better or died before returning to Paris, but suspected that he would not mind if she stayed till the old lady died! No more was said of Miles and Louisa. Sarah hoped they would soon return. It was out of her hands now. Miles and Louisa were really grown up. More grown up than we ever were, she thought. 'But you won't ever grow up, Dash, will you – lucky old dear,' she said to her spaniel. She patted his neck and then took him up to her room.

It was Louisa who insisted that Miles stay in Oxford with her as a 'husband'. Miles had thought she would take some time to get used to the idea that there was nothing now to prevent their becoming engaged to be married, but was unprepared for the almost businesslike way in which Louisa had urged this. 'Now I know that you are my best friend and that we could have been just best friends for ever,' she said. 'I want to know what else we can be. Don't you see, if we wait till we can marry it will be all spoilt. We shall be just another boring couple. I want you to be my lover before you are my husband.'

He was amazed, but how could he refuse her? They needed a few weeks rather than a few days together, but that was impossible. 'I want to go straight away to your mother with the news,' he said. 'She must not think of me as a seducer.'

'You are not one,' said Louisa, stroking his hair as they sat in the small bedroom of a lodging house off Walton Street where they had represented themselves as a married couple. 'Anyway,' said Louisa, 'René Boissier wanted to seduce me last year. I was so miserable I nearly let him. Now I have told you everything.'

'René Boissier! Good grief, Louisa – I'm not exactly surprised – but I'm relieved you didn't let him. If you are quite sure that that can be *my* privilege.'

'Quite sure.'

Their lovemaking was fierce and passionate and even tearful. There was so much they had to forget. Yet they did forget it and at the end of three nights they knew that their passion was not only mutual but, they felt, pre-ordained. There was something mysterious in the way Louisa felt that he was like her Papa, whom she had always loved, and in the way that Miles felt she was, in her rashness and vivacity, something of what Guy had once been. There were no holds barred. The soupçon of a relationship once thought incestuous even goaded them on to a sort of wildness. Yet Louisa knew that Miles was not wild and he knew that she was sincere.

'I shan't return to my studies in Oxford,' she said sleepily on their third night. 'It would be an anticlimax. I will come to America with you.'

'No, I've decided to stay on in London. I can finish my studies there as well as anywhere. And you could go to that place in London your nice uncle mentioned – if they will have a married lady as a student? Tomorrow I shall buy you a ring – it can't be too dear, as my allowance has nearly run out, but I insist.'

They were not always solemn, indeed seemed to live in an atmosphere of such happiness that jokes and silliness found their natural place. But *au fond* they were a serious couple and knew they must make their peace with the world. They had dared to rebel against its conventions and felt the better for it. The whole of their lives was now before them and they talked of their work and future as companionable conspirators.

'We'll go back to Sarah and get our things and then on to Langthwaite. Shall you want a wedding in church?' she asked.

'If you do.'

She thought it over. 'It would please the villagers, I suppose. Really I'd prefer the Embassy in Paris. What do you want?'

'Whatever you like. It's usually the bride who chooses, isn't it?'

'They'll be wondering where we are, I suppose,' said Louisa. 'We shall have to return tomorrow. This was our real honeymoon though, wasn't it?'

'Dear, delicious girl – yes, yes and yes. Paradise. We are so lucky.'

They took the train back to Bancote and arrived at Lake Hall Park first before going on to Westmorland. Louisa sported a tiny pearl ring on the correct finger and even allowed one of Sarah's maids to accompany them north.

'What a pity you just missed seeing my husband,' was all Sarah had said, too glad for them to be angry at their flight. Unchaperoned, staying together, behaving like the lower classes! But she forgave them. Louisa was just like Guy Demaine in that one respect, that she enjoyed the present moment; and unlike her mother, she had chosen to enjoy it with the right man. She tried not to think of Ned and Kate all those years ago. Marianne's letter had given her the details.

'We shall be married as soon as possible,' Miles assured her as they left.

'They are in love?' said Olivia to her mother when they had gone, as they sat sewing and reading together the evening after their departure.

'Yes, dear, he is a fine young man – '

'And Louisa is very pretty – he is lucky,' said Olivia. 'I expect they will have lots of children. I don't think *I* shall ever marry, Mama.'

'Why, dear, of course you will,' replied Sarah.

Eight months after her parents' marriage Alice Demaine was born at High Pines, a strong little baby.

Guy saw both his daughter and his granddaughter at her christening, the first unknowingly, the second with pleasure. Louisa took the meeting coolly, only afterwards confessing to Miles that she had felt really strange. 'I think he is the sort of man who takes risks,' she said. 'But risks are not much for men, are they?'

'We shall keep that secret to the end of our lives,' said Miles. 'Now we have Alice, she has reunited what was split apart. I thought my Mama seemed pleased, did she not?'

'Not half as pleased as mine,' said Louisa.

They looked out of the window. It was spring again and the wind that swept round the lake was tossing daffodils in Wordsworthian hosts in every field near the shore. The other guests had gone and they could see Marianne wrapped well against the wind, cradling their precious infant in her arms as she walked in the garden. The faint scent of Lent lilies, as they called them in that part of the country, wafted in. Bright and seemingly delicate, but amazingly tough and strong, they were in their borrowed gold, ready every year to bloom again from their coffined bulbs.

Marianne was half crooning, half chatting to Alice as she held her up to look at the hills with the daffodils. 'My sweet – I expect *you* will have secrets one day, but there is plenty of time for that.' All that misery and heartbreak, she thought, and all that love –

and Alice was the end result. She loved the child passionately, the child who had her own blood and Ned's, and Guy's.

'What are you saying to your granddaughter?' asked Sarah, who had stayed on for a few days and came out smiling into the garden.

'That I hope little Alice will be like her godmother, Sarah,' answered Marianne.